Gravity's Mysteries

From Ether to Dark Matter

Order this book online at **www.trafford.com**
or email orders@trafford.com

Most Trafford titles are also available at major online book retailers.

Printed in the United States of America.

ISBN: 978-1-4669-1017-1 (sc)
ISBN: 978-1-4669-1019-5 (hc)
ISBN: 978-1-4669-1018-8 (e)

Library of Congress Control Number: 2012900932

Trafford rev. 02/06/2012

 www.trafford.com

North America & International
toll-free: 1 888 232 4444 (USA & Canada)
phone: 250 383 6864 ♦ fax: 812 355 4082

To those who defy it

Contents

Acknowledgments

Thanks are due to my colleague Thomas Flöck, member of the German Amateur Astronomy Association 'Sternwarte Feuerstein', for his careful review of my astronomical details and valuable recommendations.

I also thank Olivier Schreiber and Gene Poole who were enthusiastic first readers of my earlier books of the same genre. Their comments and corrections were again instrumental in the final outcome of this writing.

Attila Jeney deserves my gratitude for his expertise in producing quality figures. The photos of gravity's heroes are from various public domain archives.

I also appreciate the contributions of the staff at Trafford Publishing, especially the support of Nika Corales coordinator.

First new Moon and high tide of 2012

Louis Komzsik

Prologue

Gravity. It is the most detectable physical phenomenon of the universe. Nothing or nobody can escape its grasp even at very large, astronomical distances. It is manifested in many aspects of our lives. Yet, we are still not really sure how does it work.

This book grew from readers' interest in my earlier book about the rotational phenomenon titled "Wheels in the Sky". The chapter dealing with the celestial motion of planets and the related discussion of gravity and rotational forces brought several reader questions. Answering those gave me the idea that the gravity topic itself is worthy of a book.

The book is a historical journey through millennia of encountering and gradually understanding gravity's mysteries. It starts with the early interpretation of gravity by Aristotle that was more philosophical than physical, then proceeds through the emergence of understanding by Galileo, to the Newtonian theory, that is by and large still acceptable to describe most everyday phenomena of our life. These first three chapters take us through about 2,000 years until the end of the 17th century.

They are followed by resolving some of the early mysteries of gravity, its effect on Earth's oceans and on the behavior of light, as well measuring its accurate value. These topics also have their prominent pioneers in Laplace, Eötvös, Michelson and Eddington, roughly in the 18th, 19th and 20th century.

Then we reach Einstein's theory of gravity apparently explaining the discrepancy in Mercury's precession that led to the displacement of Newton's theory. This, however, still did not turn out to be an all encompassing theory. While it led to intriguing consequences that were proven by experiments, it also left a few unexplained mysteries. Whether those will lead to another theory in the future is not known yet, but certainly interesting to contemplate.

The topic of gravity brings the opportunity to embark on side trips questioning some prevailing scientific beliefs. There are controversies about the validity of the experiments disproving the presence of ether, a topic of contention for 2,500 years and about the proof of the expanding universe hypothesis. Most of these side trips would lead to a theoretical discussion beyond the scope of this book.

Therefore these will be briefly mentioned in the chapters where appropriate, but will not be followed through in order to avoid digressing from the main topic. This carefully chosen path of the book is also maintained to keep the attention of the intended audience, the everyday reader.

1

Natural motions

The leader of the hunters spread the group around the trees surrounding the clearing where the animals grazed. He looked ahead into the distance where the edge of the clearing was, beyond which the terrain dropped down suddenly. He reinforced the flanks of his strike force and gave the sign. The group suddenly jumped up, lances raised, yelling and ran toward the animals. The reaction was immediate, the huge animals turned in the opposite direction and stampeded to the edge. One or two of the closest animals could not stop their forward momentum and fell, while the others steered sideways away from the danger and ran along the cliff to safety.

The ancient hunters carefully descended on the cliff side and mercifully killed the animals that were still moving. The huge mammoths were the secret to the survival of their tribe during the coming winter. Their meat would provide the food, their hides some clothes and even their bones would be used for a variety of purposes.

Objects falling down must have been a mystery for early humans. That they recognized the inevitable nature of the phenomenon is clear since they took shelter in caves from things falling from the sky. They also exploited the phenomenon as the above scene and other archeological evidence suggest. The observation and its exploitation was satisfactory for the early man.

Objects attracting each other were also known for millennia. The Greek mathematician, Thales (of his circle theory fame) of Miletus had recorded observations about the attractive force between amber and hay, after rubbing the amber with a soft cloth. This is of course the well known electrostatic phenomenon we sometimes notice between a pair of socks and other pieces of clothing just out of the dryer. In fact the word electron derives from the Greek word for amber.

Thales also described the attraction of iron to certain natural rocks. Those were obviously naturally magnetic, hence it is safe to say that the magnetic phenomenon was also known about two and a half thousand years ago, considering that Thales lived from 547 BC to 624 BC.

The first scientific explanation of the falling phenomenon is credited to the Greek philosopher Aristotle. Aristotle was born in 384 BC near Thessaloniki, already a city then and still striving today. He went to Athens to study in Plato's school when he turned 18 and spent another 18 years of his life under Plato's tutelage. After Plato's death in 348 BC, not being named the successor at the Academy, as Plato's school was called, he left Athens and traveled extensively while researching the flora and fauna of various Greek islands.

Half a decade later he was appointed to be the teacher of the future Macedonian King, Alexander the Great.

Aristotle

After teaching and advising the young Alexander for 8 years, he returned to Athens and founded his own academy, called Lyceum. He wrote most of his scientific work in this school during the next dozen years of his life. One of his publications was titled "Physics" and this is where the subject of the falling phenomenon was first discussed in detail.

Aristotle based his interpretation on the then commonly accepted view of the universe consisting of four major elements: earth, water, air and fire. His philosophically founded theory was that all motion in life is either natural or initiated (a loose translation of his word). Initiated motions were like throwing a rock or a spear, or a horse jumping. This is already a rather clever distinction since it separates the unexplained source and uninitiated nature of things falling.

He also prioritized between them and said that the natural motion will ultimately always subdue the initiated motion. For example the rock thrown straight up by a human will ultimately fall back to Earth, the natural motion will always win. This was clearly based on observation of real life events and it was a very credible explanation of the phenomenon.

Since Aristotle was also heavily involved in philosophy, in fact nowadays he is better known as a philosopher than a physicist, he was contemplating the causes of the two kinds of motions. He considered two types of causation: intentional cause for the initiated motion and spontaneous cause for the natural motion.

One of his strongest beliefs was that there is no motion without a cause. In the case of initiated motions, the cause was a force. The speed achieved by the object initiated by the force was proportional to the force. He even attempted to quantify this, a genuinely advanced thinking of his time, by stating an equation: force = resistance · speed.

This was based on the observation that a body will travel faster in a thinner medium, like air, than in a denser medium, like water. The topic of quantifying the resistance was somewhat difficult and carried in it a contradiction. It implied that when the resistance was infinitely small, the velocity by a finite force would have to be infinitely large. This forced him to conclude that there is no such thing as vacuum, or space without any resistance. He had to fill his empty space with something that he called ether and used later on to help him to explain another observation.

Having dealt with the initiated motions, he turned his attention to explain the natural motions. His argument was that natural motions occur along straight lines of the universe and their goal is to reconnect the four major elements. The natural lines, adhering to the Earth centric philosophy of the times, were directly down toward the center of Earth for earth-like materials or directly up for air-like materials. This was in full agreement with the observation of heavy things dropping, but air and smoke rising.

He even extended this to water: small creeks flowing into bigger ones, then into rivers and finally the one ocean of Earth. How fire fit the philosophy is not clear from the historical records, but one could make the case that many small fires will likely result in a very big fire.

Aristotle focused more on earthly things falling. He made an extremely important step in his explanation of the phenomenon when he made a distinction between the speed of similar events. He proposed that the heavier objects, made more of earth than air, were falling faster than lighter objects. This, as we now know, was a fallacy, but was plausible based on comparisons between a stone and a feather.

Aristotle also correctly observed that a falling object increases its velocity during the fall. He just incorrectly concluded that an object twice as heavy as

another one will fall twice as fast. This was mainly due
to the fact that the falling phenomenon was fast and
the differences were immeasurable by the time measuring devices of the time. It took Galileo's genius to
devise a way to slow down the phenomenon to be able
to measure it as we will see it later.

Aristotle also contemplated a scenario, now known
as his paradox. He considered the case of two stones
being on top of each other falling. He assumed that
the upper stone is pushing the lower stone and that
will fall faster. This is of course false, both stones are
falling under the same acceleration of gravity.

There were already challenges to Aristotle's weight
based hypothesis in antiquity. One notable of those
was by the Roman engineer Vitruvius who lived from
80 BC to 15 AD and who was a prolific scientific
writer. He wrote ten books on various subjects and
invented many things, including central heating. Yes,
two thousand years ago Romans built buildings with
centrally delivered heating, proven by architectural remains. But we should not be surprised; after all, aque-
ducts built by them are still standing and in some cases
still being used.

Vitruvius challenged Aristotle's view and said that
the falling behavior does not depend on the weight.
He is credited with the following counter example: If
one puts a big rock on the surface of a barrel of quick-
silver (mercury), it will float on the surface. On the
converse, if one puts a small pebble on the surface of
water, it will submerge.

He drew the consequence that falling depended on the nature of the substances involved. This was on the right track in its recognition of the weight not being the exact cause, albeit the conclusion was based on a premise involving a third intermediary object, the fluid in this case.

But it was a correct observation of Aristotle that the medium in which the object is falling influences the speed of the event. He noticed the difference between the speed of a rock falling in air as opposed to in water. This is in a sense still valid, although as we know now, the difference is between the resistance exerted by the media, not in the difference in the phenomenon.

Combining all these with the ultimate philosophical aspect that we humans be of earth and will become of earth (from dust to dust), our bodies' falling was also explained. All in all, Aristotle's theory of the falling phenomenon was in great agreement with observations and philosophy, and was widely accepted.

Aristotle's theory was, however, in contradiction with the way celestial objects behaved since Moon and the stars were apparently not falling. Aristotle would not have been the genius of his time, had he not been driven to find some explanation. And he did. He proclaimed that the celestial objects move in a different medium than the four earthly ones and as such were not subject to the same motion laws of Earth.

He actually called this medium the fifth element, besides the four basic elements: earth, water, air and fire. Since these were also called essentials, Aristotle was probably the originator of the phrase quintessential by calling this to be the quinta (as in fifth) essentia. That heavenly medium was his ether and he stated that all heavenly bodies are moving in it while obeying some divine rule that we do not comprehend.

Aristotle was an avid inquirer about nature, including the behavior of light, a topic of utmost importance in later chapters and a subject of scientific debate for most of human history. In fact he already proposed that light was a wave-like phenomenon. These were not his words, but his proposition that light is a disturbance carried by the elements of air was rather insightful, if not fully accurate.

The fact that Aristotle was wrong on the role of weight in the falling phenomenon and, according to the current scientific beliefs, wrong on the presence of ether does not diminish his extraordinary stature as one of humankind's best thinkers. Aristotle's theories were highly plausible explanations for his observations.

It is generally true that theories in any science are explanations of observations. Until the increased fidelity of the observed data compared to the actual physical phenomenon refutes a theory, it must be respected.

Once experimental technology advanced to the level to be able to demonstrate that two objects with different weight fall in the same time, Aristotle's theory did not fit the observation and had to be replaced. This, however, happened only about two thousand years later.

2

Rolling balls

In order to dispute Aristotle's assumption that weight caused the increase in the speed of falling objects one would have had to be able to show by experiment that this was incorrect. Since the problem was the high speed of the phenomenon, Galileo decided to slow it down.

Galileo Galilei, was born in Pisa in 1564 and was also educated there, except for a few teenage years in Florence when his family lived in that town. He undertook medical studies at the university in Pisa obeying his father's request, but studied mathematics for his own interest. Ultimately he even received a professorship of mathematics at the university. After only three years there he moved to the university of Padua for two decades of scientific work during which he made many of his now famous observations.

Galileo was contemplating a body moving on a slope under the force of gravity alone. Galileo recognized the fact that the horizontal and vertical components of the motion can be separated. We now call this the superposition concept, but that was not known at the time. He devised an experiment on a slightly inclined wooden board. We don't know the exact dimensions of his board, but the inclination was relatively small compared to the length of it as shown in the figure.

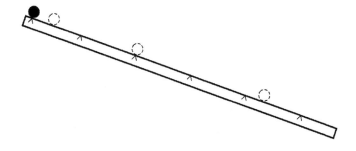

The pace of the ball rolling down on the slope was measured by a time keeping device. It could have been a water clock, a practical device to measure the speedy event of the experiment. The water clock was a vessel filled with water and had a hole through which the water drained. The level of water in the vessel, usually observed with the aid of a floating marker, was indicative of the time passed. The resolution of the clock was controlled by the size of the hole that could be adjusted.

It could also have been a pendulum type device, as Galileo was the first to propose such. It has been recorded that he observed the lanterns hanging from the ceiling of his church swinging evenly, independently of their weights. The period of the swing only depended on the length of the rope. Whether this observation influenced his thinking about the uniform effect of gravity is not known. It is, however, very clear that his science was also based on observations just like Aristotle's. What Galileo brought to the science was his pioneering ability to mathematically express experimental findings in equations.

The time keeping device needed a fast pace, probably about a second between time periods, enabling Galileo to follow the path of the rolling ball. The location of the ball was marked with a chalk at the end of each time period. After completing the experiment, Galileo compared the distances between the marks. He found that the distances covered by the ball at the end of the second period was 4 times the distance covered in the first period. The distance in the third period was 9 times that of the first and in the fourth period 16 times longer.

He concluded that the motion due to gravity was increasing its distance proportionally to the square of the time, since the distances were $4 = 2^2, 9 = 3^2, 16 = 4^2$ units and so on. It is known from his writings that he did not use the modern word gravity, he used something that loosely translates into heaviness. He wrote about the heaviness of objects, but from now on we will be using the modern terminology.

Galileo executed rolling experiments with balls made of different materials and the results did not change. He conjectured therefore that the rolling balls have the same constant acceleration produced by gravity. Galileo also understood that the time required for a body to reach a certain velocity from stationary position with a constant acceleration is the same as the

Galileo Galilei

body would need to travel the same distance with half of the final velocity.

Combining the two observations, he arrived at the value of the acceleration of gravity. Galileo succeeded to mathematically describe the phenomenon he measured. Galileo did not use the g notation but we will use this as it is the current standard. There are also some who argue that the experiment could not have produced accurate enough results due to lack of accurate time keeping devices of his time. Nevertheless, some records indicate that he was very close to 10 meters per second squared. We now know that g is 9.81 m/sec^2 in metric and 32.2 feet/sec^2 in English units.

Whether his rolling experiments enabled him to calculate an accurate value of the acceleration is immaterial. The fact that a constant value was at work was the important recognition. Galileo considered the constant to be the same anywhere on the surface of Earth, but we will see in the following chapters that there are some variations, some of them rather intriguing.

Then Galileo concluded that when the cause of the acceleration is exclusively gravity, like in the case of falling bodies, the loss of height of the falling body is computed by the exact same formula. This formula is now known in even first grade high school physics as the equation of free fall.

According to a sometimes disputed story, Galileo in the early 1600's also executed an experiment at the leaning tower of Pisa. He demonstrated that a wooden ball with the same shape as an iron ball reaches the ground at the same time, when dropped from the high tower. This proved that the acceleration of a body in free fall is constant, independent of its weight, assuming the shape of the body does not generate a measurable amount of air resistance.

This story may be false and the event may have never happened. Clearly it was not the experiment that led him to his equation. It may have been though actually executed to demonstrate his findings to a lay audience later. Or, it may have been a duplication of an experiment he heard about. Simon Stevin, a Dutch scientist allegedly executed the same experiment at the church tower of Delft, some twenty years earlier. Galileo was not opposed to follow other people's ideas, after all it is

well known that he built his first telescope after hear-
ing about such built by another scientist, incidentally
also a Dutchman. Whichever was the real cause we
will never know, one thing is certain; the Pisa tower
experiment of Galileo is now forever embedded in the
history of science.

Galileo was also interested in the motion of projectiles. He again employed the superposition concept
and decomposed the motion into horizontal and vertical components. He assumed that the distance traveled horizontally is simply proportional to the velocity.
By that time this distance versus velocity relationship
was well known, since even everyday persons knew that
a faster horse travels a longer distance during a day
than a slower animal.

Galileo further understood the concept of inertia,
even if he did not call it such. He knew that a body
would continue with a constant speed until its motion
is modified by a force. If an object is thrown horizontally in an environment without any friction or resistance, it will move with a constant velocity indefinitely.

On the other hand the object will undergo the same
motion vertically as if it were just dropped and he was
the one who derived the formula for that. Galileo recognized the fact that the time is common between the
horizontal and vertical motions. By equating the time
of the two motions he proved that the shape of the
trajectory is the well known parabola. That meant
that the height change of the projectile was also proportional to the square of the horizontal distance, assuming that the effect of the air resistance is negligible.

Galileo therefore could compute the maximum height and the maximum distance attained by any projectile. His knowledge soon found its way into military science and became the foundation of ballistics. It was applied to calculate trajectories of cannon balls and other projectiles of warfare.

Examining the case when the projectile is shot directly upwards, he realized that the motion due to the initial velocity is countered by gravity. Once the velocity produced by the acceleration of gravity is the same as the initial velocity, the object will reach the top of its trajectory. It will then fall back again and will hit Earth with velocity equal to the initial velocity. Hence Galileo recognized the potential energy manifested in the elevation of an object. He was a step away from establishing the concept of the gravity field, a topic of much importance later.

Galileo's superposition concept leads to some interesting consequences. If a bullet is shot out horizontally from a high powered rifle with a very high velocity and is shaped for minimal air resistance, it will go a long way before it drops on the ground. On the other hand, the time transpired during the flight of the bullet is the same as the time it takes a bullet to fall to the ground when dropped from the height of the rifle.

This is a somewhat counter-intuitive scenario, often exploited by manufacturers of firearms who brag about their rifles: It hits the target 1500 feet away in about the same time as it takes for another bullet to fall to the ground from 5 feet. But we know from the superposition principle that this is somewhat of a false claim as the effect is really due to gravity. Some credit of course may be given to the manufacturer for designing a bullet that minimizes air resistance and getting as close to the theoretical result as possible.

As revolutionary as Galileo's thinking was in his time, he was not infallible. He was asked by a high ranking member of the Catholic church to explain the tides. He attributed them to the rotation of Earth around its axis and the resulting sloshing of the ocean's waters. This of course contradicted the observation of two daily tides on the Adriatic coast of Italy, specifically in Venice where the topic was of immense interest. He dismissed this as a result of the local behavior of the Adriatic, due to its special shape and semi-enclosed nature. He was wrong and the phenomenon is of course due to the gravity pull of Moon as we will see later.

There are also written records indicating that Galileo was attempting to compute the center of gravity in solids. That could have been the first step into the direction to understand that the motion of planets is also governed but the force of gravity, but he did not reach that. This spectacular achievement was accomplished by another genius born within a year of Galileo's death. That person was Galileo's immediate successor in solving the mysteries of gravity, Newton, who started to focus on the Moon but then traveled far beyond.

3

Dropping apples

Isaac Newton was born on a farm in Lincolnshire, England in the year of 1642. He was a prodigious student and enrolled at Cambridge University soon after finishing his grade school studies. In the late 1660s, during a cholera epidemic in London, he returned to the family farm for about a year. According to a now enshrined anecdote, possibly as fictitious as Galileo's Pisa story, one evening having tea in the apple orchard behind the family house he noticed an apple dropping and hitting the ground.

Newton realized that the force of gravity was pulling the apple to the ground and started to contemplate how far the force reaches above the apple tree. He concluded that it is likely to reach the Moon. He extended this thinking and considered that even Earth itself is pulled by a similar force, that of Sun's. He wanted to quantify these relationships and needed to find out how this attraction force changed with distance.

According to historical records, the idea of the force of gravity changing by the distance squared was originally proposed by Robert Hooke, the inventor of the law of springs. Hooke was, however, an experimental scientist without the mathematical acumen of Newton

Isaac Newton

who ultimately got the credit for it by actually proving the correctness of the conjecture. This resulted in a lifetime of animosity between the two of them, but not the only one of such in Newton's life. He was apparently rather selfish and ruthless in such circumstances, but we will leave that topic for the biographies. His brilliant insights into gravity are our subject.

The way Newton came about proving the conjecture is worthy of discussion. The radius of Earth and the orbit of the Moon were well established by Newton's time and from this he calculated the length of the orbit of the Moon as the circumference of a circle. The time it takes the Moon to do one circle, the length of a lunar month, was also well known. He then computed the orbital velocity of Moon by dividing the circumference of Moon's circle with the time Moon takes to complete it.

On the other hand Newton concluded that if on Earth an apple falls about 16.1 feet per second (since the acceleration of gravity is 32.2 and the free fall formula of Galileo uses half of that) then the apple (or any other object for that matter) falling from Moon's orbit (that is about 60 times farther than the radius of Earth) would fall 1/3600th (3600 being 60^2) of 16.1 feet in one second. This is equally true for the apple or the Moon.

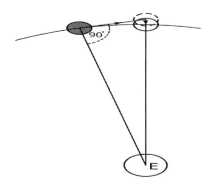

Newton sketched the scenario shown in the figure above. The arrow tangential to the orbit represents the direction Moon would travel without gravity due to its orbital velocity. The length of the arrow is the distance Moon would travel in one second. The smaller arrow pointing toward Earth is the amount Moon would fall due to gravity.

He understood that if Earth's gravitational pull would be removed, Moon would continue uninterrupted along a tangential direction to its orbit. On the contrary, if Moon became stationary by losing its orbital velocity, it would fall back to Earth. The continuing pull of Earth keeps Moon in orbit by its direction pointing always toward Earth independently of the position of Moon on its orbit.

Using the Pythagorean theorem in the right triangle of the figure Newton was able to verify that the result was astonishingly close to the expected value. Newton therefore proved in connection with Earth and Moon that gravity is diminishing by the square of the distance; a spectacular feat of logical thinking.

Newton went back to Cambridge and spent years refining the theory based on his lunar calculations. His final law of gravitation states that two masses will pull on each other with the force that is decreasing proportionally to the square of the distance between them. The force is directly proportional with the size of the masses and a gravitational constant G. Interestingly the formula said nothing about the volume or shape of the objects, Newton considered the masses to be concentrated in their center of gravity.

Newton also conjectured that the only force acting across astronomical distances is gravity. It has the longest reach but the weakest magnitude. It is an exclusively attractive force that pulls any two objects with masses together. He was also convinced that the constant G is a universal constant, common to all objects in the solar system. This is still accepted today in the mainstream science, albeit contested by some. The value of the constant is $6.6 \cdot 10^{-11}$ m^3/(kg $\cdot$ sec^2) in the metric system.

Newton's universal gravity law applied to a body on the surface of Earth simplifies, since the distance becomes the earth's known radius and one of the masses is that of Earth. Hence the acceleration of gravity on any object on the surface of Earth becomes the value g that Galileo measured. The gravitational force of Earth on a body then becomes its weight. Since different bodies have different masses, their weight will also differ. Seems like Newton and Galileo were in perfect synchrony.

The assumption of Newton's theory that all the mass is concentrated at the center of Earth is of course physically not true and realizing this we can embark on an intriguing thought experiment. This was originally proposed by the Italian engineer Tartaglia, famous for his role in the quest for the solution of the cubic equation in the early 1500s.

Let us assume that we can drill a hole through the center of Earth, ignoring the internal crust and core and all that. Let us now imagine falling into that hole. We would be free falling in the hole, but there would be ever diminishing amount of mass underneath our feet. Assuming no air resistance in the hole, lack of friction against its wall and a uniform distribution of the mass of a perfectly spherical Earth, the Hungarian engineer Istvan Szabo in 1970 analytically computed the velocity we would reach the center of Earth with as 28,200 km/hour.

Then we would continue falling upward on the opposite side, due to the velocity we gained while falling in. The acceleration of gravity, however, would change direction and now the ever increasing mass below us would exert a force slowing us down. Ultimately we would reach the surface on the other side of Earth with zero velocity. Then we would fall back again and the process would continue with a periodic time of 1 hour and 25 minutes.

While Newton had not done this thought experiment, he certainly realized the fact that on the surface of other planets this value would be different due to the different mass of those planets. We now know for example that on the Moon, the acceleration of gravity is about 1/6th of that of the Earth, or 1.63 m/sec^2, while Sun's acceleration of gravity is 274.1 m/sec^2 or almost 30 times that of the Earth.

The application of Newton's law of gravity to the solar system enables us to compute the masses of our planetary neighbors based on their visually measured orbits. This branch of science is called gravitational astronomy and it is used to compute masses of celestial objects with surprisingly high accuracy without having ever traveled there, not that being on the surface of the planet would provide any better means to measure its mass.

In space with no air resistance a planet will retain

its orbit due to the force of gravity acting on it. In fact the law describing the orbit of a planet around the Sun, Kepler's first law, may be derived mathematically from Newton's law of gravitation, but it exceeds the intended level of mathematics in this book. The ability to derive Kepler's equations from Newton's gravity formula was considered to be the ultimate proof of the correctness of the theory.

In the case of a single planet and the Sun, the orbit derived is an ellipse. The period of the orbit depends on the size of the orbit. However, in our solar system there are more planets. The influence of the planets on each other, besides their Sun controlled motion resulted in some aberrations from the single planet orbits leading to the discovery of some of the yet invisible objects of our rotational world.

Such unexplainable motion of Uranus, for example, was attributed to the presence of another celestial body outside of its orbit. Newton's theory of gravity was able to predict the existence of a planet not known yet. The French scientist Urbain Le Verrier described the anomalous motion of Uranus with a system of 279 equations and he even solved it manually. Sure enough, the German astronomer Johann Galle in the Berlin observatory, to whom Le Verrier wrote about his solution, discovered the planet Neptune almost exactly in an orbit where Le Verrier's computations indicated it should be.

The elliptic orbit of a planet would be completely unchanging if it was the only planet around Sun. That would place the position of the planet closest to the

Sun, the perihelion, fixed. Due to the effects of the
other planets in the solar system, this position shifts
over time. This is rather normal for all planets in the
solar system, Earth also has its own perihelion shift resulting in its approximately 26,000 years long celestial
cycle, described in Cycles of Time of the Bibliography.

Such a perihelion shift, also called perihelion precession, led to the questioning of the correctness of
Newton's theory. The shift of Mercury's perihelion by
Newton's law, computed in the 1860s by Le Verrier
was 531 arc seconds (one 3600th of a degree). The
observed shift was about 574 arc seconds a century, a
difference of 43 arc seconds.

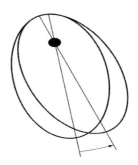

The perihelion precession of Mercury is indicated by
the arrow in the above figure, where the dot is the
Sun and the ellipses are Mercury's orbits. Assuming
the absolute validity of Newton's law of gravity, Le

Verrier was on the hunt again. He was so much convinced that the law was inviolable and there must be a planet that he even named it in advance to Vulcan. Despite some false findings, however, the planet was never found. This opened up the room for another theory of gravitation leading to Einstein.

Einstein's theory, subject of a later chapter, was able to account for the 43 arc seconds discrepancy, hence it was considered to be the new "rule of the universe". But Einstein's theory now also has difficulties accounting for some observed phenomena, discussed later. Hence there are now new efforts to rehabilitate Newton's law.

One of the possibilities is to allow the gravitational constant of Newton, G, to be varied. It could vary in distances very close to or very far from large gravitational masses, after all, we only have one place to measure it in the solar system. If it would be somewhat different in the close proximity of the Sun it could even explain the Mercury anomaly. Due to the scientific entrenchment of Einstein's theory, such efforts are highly disputed and we will not follow them.

There are also some voices in the scientific community stating that Einstein's proving of the anomaly is actually false. They propose that the calculations of Einstein were actually based on incorrect observations of the rather infrequent phenomenon. Mercury's transit occurs sporadically in a pair of 3 year events and the most recent pair was in 2003 and 2006. Clearly the earlier comparison calculations were also based on the 3 year pairs.

It is, however, now recognized that one of the pair is always in May and the other is in November, and that poses a problem. In May Earth is below the ecliptic and in November it is above, and since the observation of Mercury's perihelion is from Earth, the three year pairs are actually not comparable. The paths of Mercury leading up to and following the transit have distinctly different slopes, one descending and the other one ascending. It is possible that Einstein proved an incorrect observation, hence his theory does not really provide the exact result.

This topic is still undecided, since it boils down to finding the ultimate reference system in which the exact observation and calculation should be made, and that is not easy. The rotation of our solar system in the Milky Way, then our galaxy's rotation in the universe and the universe's own suspected rotation makes this road very difficult, so we abandon it.

Newton was deeply disappointed that he was not able to explain what carried the force of gravity, but he was able to explain another interesting Earthly phenomenon: the tides that flummoxed Galileo. In his work titled Principia he proposed that the gravitational forces exerted on Earth by Moon and the Sun are responsible for the phenomenon.

Newton was ultimately honored when he was entombed at the Westminster Abbey, the first non-royal person achieving that distinction.

4

Lunar tides

Imagine yourself being on the beach and watching the approach of the waterline on the shore. There is some calming and at the same time disturbing aspect of it, the wave noise could lull your senses but the relentless coming of the water makes you wonder about the forces that generated it.

The topic was subject of intense speculation throughout humankind's history, especially since the early concentrations of advanced cultures were mainly around the Mediterranean. Hence, there is a wide-spread history and many notable scientists involved in the explanation of the tidal phenomenon.

A lunar relation of tides was recognized very early, a description by the Babylonian Seleucus already attributed the tides to Moon's cycles. There are more records indicating such recognition in other sea-side cultures, for example by the Greeks and ancient Chinese, both well before Christ.

The first organized observations in the form of tidal tables appeared about a millennium later, in the 11th century in China and in the 13th century in England. Both were related to river locations in close proximity to their outlet to the sea; in China the river Qiantang and in England the Thames.

The fact that they all realized the relationship with
the Moon did not mean that they recognized the gravitational role of the Moon in it. Newton was the first
who tied gravity and tides together, and the French
scientist Pierre Laplace provided a mathematical rep-
resentation for the phenomenon.

Pierre-Simon Laplace

Laplace was born in 1749 and was sort of a successor to Newton, albeit not in such a close chronological
proximity as Newton followed Galileo. He was obeying
his father's wish in first pursuing other topics than his
life work later, a frequently occurring story in science,

another example being Galileo. Laplace's father's wish took him to study theology at the University of Caen with the goal of becoming an ordained priest. His interest, however, very soon turned to mathematics and he dropped out of divinity school to pursue it.

He ultimately received a teaching position at France's military academy in 1771 where among his students was the future emperor, Napoleon Bonaparte. He spent almost two decades there and started his masterpiece titled "Celestial Mechanics". The monumental work was completed by 1799 and published in five volumes in the first quarter of the 1800s. He dedicated a volume to the by then Emperor Napoleon. Anecdotal evidence tells the story that Napoleon complained about God not being in the volume to which Laplace replied that for the topics he discussed there was no reason to invoke God.

Laplace was a very serious student of gravity. In the above book he attributed the formation of planets and galaxies to gravity. Gravity between particles of a primordial swirling mass would cause them coalesce into bigger particles. This coalesced mass then has an increased level of gravity that may attract more particles that are still free to join. This process continues until the planets collected all the free material around their orbits, not lost in the gravitational tug of war with the neighboring planets.

The tidal effect is also result of a gravitational tug of war. Earth's acceleration of gravity on the surface of Earth is slightly modified by Moon's own gravity. The difference between Moon's effect on Earth and Moon's

effect on a body (for example the water in the oceans)
on the surface of Earth is the tidal acceleration. The
lunar tidal acceleration is about $10^{-7} \cdot g$, or about tenth
of a millionth of Earth's own acceleration of gravity.
While that seems almost negligible, it is not in the
grand scheme of things as attested by the tides we see
daily.

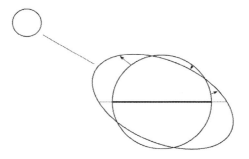

In analyzing the tides, Laplace considered Earth to
be fully covered with water. This is not as farfetched as
it sounds, after all, almost two third of Earth's surface
is covered with oceans. In the above figure the small
circle on the left represents Moon, while the large circle
represents Earth's surface. The ellipse represents the
deformed shape of the water surface of Earth, due to
the gravitational pull of Moon. The arrows represent
the direction of the moving water surface. They show
that in some areas the water level decreases, while in
others increases.

The increasing water level on the side that is closer to Moon makes sense, after all, that is where Moon's gravitational pull occurs. This side is called the sub-lunar side for obvious reasons. The other side, called antipodal, begs for an explanation. Why would be a bulge on that side also? The explanation for that is because Earth is also being pulled by Moon toward it and to a lesser extent that the water on that side, therefore there will be a higher water level there as well.

This is also the reason why we get two tides a day. One occurs as our sub-lunar tide when Moon is up and one we get as an antipodal when it is down. Tides therefore occur with a reasonable repeatability, we get high tide twice a day about 12 hours and 25 minutes apart. There is half that time between the high tide and the low ebb.

There is a strong variation of the tides depending on the relative position of Moon with respect to Earth. Specifically, the distance between them changes by as much as 50,000 kilometers during the duration of a month. Once a month, when Moon is the closest to Earth in the so-called perigee position, the tidal forces are the highest. On the other hand when Moon is at the apogee, farthest from Earth, the tidal forces are the lowest.

Another variation of the lunar tidal effect is due to the fact that Moon's orbital plane, in which it is rotating around us, is inclined about 5 degrees to the equatorial plane of Earth. Hence Moon, when making one monthly revolution, is making a roundtrip between a maximum position above the Equator to the same distance in the South. As a result, Moon crosses the equatorial plane twice a month. This effect of Moon's declination is named the diurnal tide pattern.

The tidal effect of Moon also depends on its relation with respect to the Sun. The solar tidal acceleration is about half of that of Moon. This seems contra-intuitive since the Sun is much larger than the Moon but we can prove it by simple arithmetic. Sun is almost 27 million times bigger than Moon, but it is also 390 times farther away. Since the tidal acceleration diminishes by the cube of the distance, dividing 27 million by the cube of 390 yields about one half. Sun's effect is far from being negligible and sometimes exaggerated by the relative positions of the Sun and Moon.

When Moon is either full or new, it is on the line connecting the centers of Earth and Sun. In the case of new Moon, it is on the side of the Sun and the tidal forces of both Sun and Moon are combined. These are the highest tides, sometimes called the spring tides, although they do occur in the other seasons as well. When Moon is on the opposite side of the Sun, at full Moon, the tidal forces are also combined and spring tides occur. The effect is now produced by the antipodal tide scenario, but both high tides are about the same.

When Moon is in the intermediate phases, the so called neap tides, or the low high tides occur. Their cause is that the tidal forces of the Sun and Moon are perpendicular to each other. Wherever Moon is raising the water level, the Sun is lowering and vice versa. It is not an exact cancelation, since the tidal acceleration of Sun alone could produce only about one half of Moon's tidal variations. Hence these are low high tides, but still higher than the average water level.

Finally the tidal effects also depend on the alignment of the Moon-Earth-Sun triple. Earth is about 150 million kilometers from the Sun in its aphelion position and about 5 million kilometers closer at perihelion. This is a significant difference in itself. When Moon is also in perigee (closest to Earth) at the time of Earth being in perihelion (closest to Sun), the combined gravitational effects are significantly higher. This is called the proxigee position and the resulting extra high tides are called proxigenial tides. There are even longer term tidal patterns related to certain planetary arraignments of the solar system (a notable one coming in December 21, 2012), but their million year periodicity is beyond our focus.

Laplace created a mathematical model of the tidal phenomenon as function of the average assumed water depth, and the latitude and longitude of the location. He derived a set of three partial differential equations. The three solutions of the equations were the temporal rate of change of the tidal elevation (tidal surface height), and the acceleration of tidal water in latitude and longitude directions at a certain global location.

Laplace also considered the special effects caused by Earth's rotation, the centrifugal and Coriolis forces, topics of importance in a later chapter. These are very important components of Laplace's tidal equations representing the different tidal behaviors at various latitudes, and between the Northern and Southern hemispheres.

The theoretical maximum level of lunar tides is about 21 inches. This is with the assumption of uniform depth and Earth being fully covered with oceans. The stronger spring tide could raise this to as much as 31 inches and the weaker neap tide would be about 11 inches, a variation of plus or minus ten inches. The seldom experienced special scenario of the proxigee mentioned above could produce a super high tide of 37 inches. This is still only about three feet and occurs only once in every one and a half years.

The actual water level height and tidal reach measured on coastlines is not the pure increase measurable in the middle of the ocean. Depending on the geography, when the water moves toward the shores, tidal waves occur that raise the water level well above the average. The world record of tidal elevation is in the Canadian Minas Basin in Nova Scotia. The difference between the high and low tide reaches up to 50 feet, more than 15 meters. The amount of water flowing into the basin is about 14 billion cubic kilometers and it is now established that the Nova Scotia territory tilts under this tremendous weight twice a day.

The idea of traveling to the cause of all these tides, Moon, has been in the human thinking for ages. Newton's understanding of Moon's motion and Laplace's understanding of Moon's effect on our oceans paved the way to seriously think about visiting it.

5

Moon landings

It was science fiction at the time Jules Verne wrote his famous "From the Earth to the Moon" book in 1865 that of course became reality about a hundred years later. In fact the ways of accomplishing it, by building a humongous gun and shooting a cabin containing the travelers to the Moon, was very close to the actual realization of the trip.

Let us consider the issues of such travel now from gravity's point of view. The fact that there were certain positions in the Sun-Moon system where tidal forces canceled out brings us to an extremely interesting and practically useful gravitational scenario first recognized by Lagrange.

Giuseppe Luigi Lagrangia was born in Italy in 1736 to Italian parents, but due to French heritage on his Father's side, he changed his name to Joseph Louis Lagrange. He spent half of his adult life in Berlin where he produced significant results in the area of mathematics, classical mechanics, but also luckily for our topic in celestial mechanics. Some of these results were acknowledged by the French Academy of Sciences and he moved to France in the last decade of the 1700's just in time for the French revolution.

Joseph-Louis Lagrange

During those years his paths crossed with Laplace whom we met in the last chapter, and even had a common celebrity connection in Napoleon Bonaparte. It appears that Napoleon was more appreciative of Lagrange's science, giving him the Legion of Honor and even making him a Count. Laplace had to wait for the returning Bourbons to receive his nobility (he became a Marquis), but that is for the history books. Lagrange became the first professor of mathematics at the newly opened École Polytechnique, France's premier school of higher learning.

Lagrange's analyzed the gravitational relationship between Earth and Moon. He found five locations where the gravitational forces of the two bodies are in a special relationship and fully or partially cancel each other out. These points are now called Lagrange (sometimes libration) points and their location is shown in the figure.

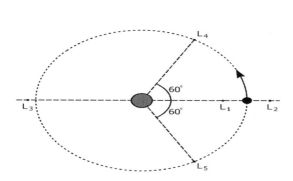

One such point is called L_1 that is the point on the line connecting the centers of Earth and Moon at the location where the gravitational forces are opposite and actually cancel each other out. The importance of that is in connection with the orbital periods. As we saw it in an earlier chapter the circumferential velocity and the gravitational force are in a delicate balance. An object located in the L_1 Lagrange point has the same orbital period as Moon.

More importantly, an object traveling from Earth to the Moon would change gravitational hosts at that point. If a spacecraft reached this position, from here Moon's gravitational pull would be its governing force. In other words, the object would simply fall onto the

Moon from just beyond this point.

But reaching that point is not an easy feat as it requires tremendous energy to lift a space vehicle off the ground. Although the velocity required for an object to escape the confines of Earth's gravity is independent of the mass of the object, the bigger the mass the more energy is needed to accelerate the object to the escape velocity.

More specifically, the escape velocity is the initial speed required from a stationary position and as such it is measured at the surface of Earth. On average, the escape velocity from Earth is about 11.2 kilometers per second. That velocity is about ten times the speed of a bullet shot from a good rifle. Clearly, we need an extremely good rifle to shoot something out of Earth's gravitational field.

The above is actually the second escape velocity. The first one is the velocity of 7.9 km/sec to put a satellite into Earth's orbit. There is also a third escape velocity of 42.1 km/sec that would enable a craft to leave our solar system. Finally, the fourth escape velocity is at 320 km/sec that we need to surpass to leave our galaxy, the Milky Way.

The value of escape velocity assumes that the object is shot up in a vertical direction from Earth and needs to fight head on against the gravitational field. Earth itself, however, could help us in this endeavor. Since Earth's angular velocity due to spinning around its axis at the Equator is about 0.465 kilometers per second, if we launch our object to the east horizontally, we only need about 10.735 kilometers per second escape velocity. It does not seem much of a difference, but when considering the energy needs of practical objects (certainly larger than a bullet), this is significant. This is the reason the launch site of our space program is as south on the continent as possible: Cape Canaveral.

Cape Canaveral was of course where the historical Moon landing expeditions launched from. The first such operations were so-called hard, and later soft landings. This distinction is due to the fact that after reaching the L_1 point the challenge becomes to reverse the gravity force of Moon for landing.

Hard landings mean essentially falling to the surface of Moon and that is somewhere in the neighborhood of 2.65 kilometers per second, some 9,500 kilometers per hour. Building a structure retaining its integrity after such a crash is very difficult. Some of the hard landings were successful in taking some photos, but the craft were destroyed.

In the second half of the 1960s several American spacecraft were successfully soft landed. The Surveyor probes were able to gather soil composition data and transfer back, electronically of course. Now the issue of actually bringing something back physically arose. Another fight with Moon's gravity came up now in the opposite direction.

Moon's escape velocity due to its smaller mass is significantly smaller than that of Earth's, but still a formidable 2.38 kilometers per second. This seems like a problem already solved, after all, if we were able to escape Earth with the space craft, escaping Moon should be easy, shouldn't it? Well, the difference is that the craft landing and ultimately taking off from Moon is not the same as the one that left Earth. As we saw in the old Apollo films and the new films of space shuttle take-offs, leaving Earth requires a multi-stage rocket whose pieces are discarded in the process.

The Moon landing craft included a dual use rocket that on arrival decelerated and on departure accelerated the craft to overcome the escape velocity. Let us now assume that we are off the surface of Moon and on a return trajectory. The fight with gravity is not over because we need to overcome again the gravity of Earth in an opposite sense and prevent the destruction of the returning capsule. That is made somewhat easier by Earth's atmosphere since it's resistance itself decelerates the craft, however, it also generates a lot of heat. Hence the landing module, the capsule containing the astronauts, contained heat shields and parachutes to complete the trip.

Moon landings with humans on board were successfully done 6 times during the Apollo program of the 1960s. Apollo 11 was the first, with Neil Armstrong being the first human to set foot on Moon on July 20, 1969. A Saturn V rocket launched from Cape Canaveral on July 16th carried the command module Columbia and lunar landing craft Eagle into orbit. Columbia was actually named after Jules Verne's giant cannon.

Neil Armstrong's first steps on an extraterrestrial surface were watched by more than half a billion people worldwide. The extraordinary feat was repeated by Apollo 12, 14, 15, 16 and 17 with the last human on Moon being Gene Cernan on December 11th, 1972. Apollo 13 was of course the so-called "successful failure" mission commemorated in the movie of the same title. The crew was forced to take refuge in the landing module due to an onboard explosion and returned safely to Earth after taking a loop around Moon.

Returning to the Lagrange points, the second point, L_2 lies on the same line but just a bit outside of the orbit of Moon albeit the figure does not show it. The Earth-Moon L_2 point could be a good location for a space based observatory to see Moon's far side. The third point, L_3 is on the opposite side of Earth. Finally L_4 and L_5 are located at 60 degrees from the line connecting the two bodies and again slightly outside of the orbit of Moon.

Naturally, Lagrange points exist in connection with any pair of celestial objects. The practical importance of such locations is enormous. For example, the Sun-Earth L_1 point is a prime location for space based observatories of the Sun. In fact that is where the SOHO satellite (Solar and Heliospheric Observatory) is located.

The L_2 of the Sun-Earth system is also a good location for a space based observatory. but now looking away from the Sun. It also has the desirable characteristic of having the same orbital period as Earth and it is the location of the Kepler space telescope launched in March 2009. The telescope's first success was in 2011 by the finding a star with a habitable planet, now named Kepler 22B about 600 light years away.

The Sun-Earth system's L_3 is on the other side of the Sun and somewhat less interesting for humankind. That is also less desirable for practical use because the gravitational effects of the other planets will come into consideration and make it instable. It was part of the science fiction lore for some years as the possible location of the twin of Earth's, an appealing concept to lovers of symmetry theories.

The L_4 and L_5 points' balance is even more interesting. As indicated in the figure above they are forming an equilateral triangle with the two celestial bodies of the system. The balance there is such that the ratio of the gravitational forces from the two objects is the same as the ratio of their masses. The objects placed in these points will be in synchrony with the system. The L_4 point is orbiting 60 degrees ahead, the L_5 point is 60 degrees behind the smaller body, almost like a celestial front and a rear guard.

These points are very stable and can contain objects with relatively large mass. Nature itself recognized this fact. The Sun-Jupiter system's L_4 and L_5 points are occupied by two asteroids, called the Greek and Trojan asteroids, respectively. Hence sometimes those points are also called the Trojan points.

A recent Moon related experiment, albeit not resulting in landing, was the September 2011 launch of a pair of spacecraft from Cape Canaveral atop Delta II rockets. The mission was historic in the sense as these were the last of such rockets to be launched.

The importance of the mission for our topic is, however, related to its goal: executing gravitational measurements in hopes of finding out the interior composition of Moon. As one of the NASA program managers said "this is a journey to the center of the Moon". The mission is looking to find out whether Moon has an inner core. Does it have fluid around it and an Earth-like mantle, or is it just a big piece of round rock?

The mission is called GRAIL (Gravity Recovery And Interior Laboratory) and the craft were aptly named by a group of 5th graders in the elementary school of Bozeman, Montana to Ebb and Flow. The craft used a special trajectory that took them first to the neighborhood of the Sun-Earth L_1 point that is about a million miles from Earth. The advantage of that trajectory was in its lower energy requirement to escape Earth's gravity but it took longer, almost about 3 months to reach Moon. In contrast, the Apollo missions reached Moon in only about 3 days.

The craft reached Moon's orbit on New Year's day in 2012. After some adjustments they will be in a circular orbit with a radius of approximately 30 miles and an orbital period of about 2 hours. They will orbit in tandem while maintaining their distance to an accuracy of less than ten thousandth of an inch.

Since the energy of the craft is provided by solar panels, it is unclear how long they will retain their orbits. There is a lunar eclipse coming in June of 2012 and whether the craft's solar components survive that celestial positioning remains to be seen.

As we learned from Newton, Moon is in a delicate balance with Earth's gravitational force and its orbital velocity. The same balance will be true between Moon and its own two little "moons", GRAIL A and B. During their orbits, whenever the gravitational force changes their orbital elevation will also change. When the gravitational force becomes lower, the craft will be elevated to a slightly higher orbit. On the contrary, when the force of gravity is higher, they will be somewhat lowered.

Since they are in a distance of 120 miles from each other, they will encounter such gravitational fluctuations at different times and that will enable accurate relative measurements. These variations will provide information about the interior constitution of Moon.

Potentially the mystery of Moon's creation may also be resolved. The prevailing scientific belief is that Moon was created when another planet crashed into Earth. The material ejected by the impact gradually coalesced into bigger and bigger pieces, ultimately becoming the Moon. There are, however, recent hypotheses about another body also slamming into Moon, resulting in the rough and highly scarred far side of the Moon.

The remaining question is: why would the force of gravity vary with local features of Moon? Would that also be valid for Earth? The resolution of this mystery is next.

6

Earthly masses

There was intense scientific interest in measuring the accurate value of the acceleration of gravity during the 19th century and there were many attempts. Among them was an ambitious expedition by a group of German scientists on board a ship on the North Sea sometime in the 1860s. To their consternation the results of their gravity measurements, while very repeatable, varied between the East bound and West bound paths of the ship. This was not understood, but reported in their published documents of the experiments.

The mystery was solved by a Hungarian scientist, Baron Lórand Eötvös, who received his doctorate from the German University of Heidelberg in 1870, a noted scientific stronghold of the days. Eötvös was born in Hungary in 1848 of a noble family, received an excellent education afforded by the family's wealth and of course exploited by his superior intellect. By 1871 he was a full Professor at the University of Sciences in Budapest, now named after him, and in 1873 became a candidate member of the Academy of Sciences. He later became its President, even a Minister of Education in the Hungarian government and had a very distinguished scientific and political career. International accomplishments followed, including the Legion of Honor, France's highest award that Lagrange had also received, and even a crater on the Moon was named after him. His immortal fame is, however, intrinsically linked to gravity.

Loránd Eötvös

Eötvös received a copy of the German scientific report about the variation of acceleration of gravity depending on the direction of travel and he instinctively knew the reason for it. He realized the fact that Earth's rotation is at work. More specifically, the by then known Coriolis force, occurring in any rotational system is the culprit. Coriolis force is now widely known to be the cause of many interesting Earthly phenomena, such as hurricanes and trade winds, but not yet connected to gravity at that time.

Coriolis force was then understood to be acting in a rotating plane with an axis of rotation perpendicular to the plane. Eötvös realized that in the case of Earth, this was really only true at the poles. At any other latitude, the plane of rotation is at an angle with respect to the axis of the rotation (i.e. Earth's), but the Coriolis force acts in a plane perpendicular to the rotation axis. Hence at a certain latitude, there is a component of the force that is perpendicular to the local horizontal plane. This component of the force now is called Eötvös effect to honor his recognition.

An interesting aspect of the Coriolis force is in its orientation being always in a strict relationship with respect to the axis of rotation and the movement of the body experiencing it. A body moving in the opposite direction, or by the same token under an opposite rotational direction, would detect an opposite Coriolis force orientation. The opposite rotational direction accounts for the differences in the weather phenomena between the Northern and Southern hemispheres.

What happened to the gravity measurements of the German scientists was the first case. The opposite direction of the movement resulted in opposite orientation of the Coriolis force. That meant that its vertical component, the Eötvös effect, was pointing downward, hence strengthening the gravity force, when moving west and pointing upward, hence decreasing the force of gravity when moving east. Eötvös was ultimately vindicated by a repeated set of tests on the Black Sea that produced the same discrepancy between east

and west bound measurements. That proved that the
North Sea experiment was not a fluke and Eötvös' explanation was correct.

Having resolved one gravity mystery, Eötvös continued to investigate the effect of the rotation of Earth
on gravity. He further recognized the fact that the
centrifugal force of Earth's rotation makes the direction of the force of gravity vary with latitude location.
This variation is the highest at 45 degrees latitude and
Budapest was located at 47.25, a perfect location to
do some experiments. According to Eötvös' calculations, the angle of difference between the theoretical
and actual gravity direction in Budapest is about 356
arc second, a small but not negligible, and according
to his thinking, measurable angle.

He set out to measure this effect and devised a very
simple, but ingenious equipment he called the curvature variometer. It was simply two balls at the ends
of a horizontal rod suspended at its middle point on
a string. He knew that the balls were simultaneously
under the centrifugal force, pointing sideways, and the
force of gravity pointing to the center of Earth.

The resulting force was a combination of both forces.
He assumed that the balls were close enough with respect to the radius of Earth that they were considered
to be on identical location. Eötvös hypothesized that
the centrifugal force is independent of the type of the
material, but if the force of gravity would depend on
the type of material, the resulting force would be different on two balls made of different materials of the
same mass. He was aiming to measure this horizontal

force difference if it existed.

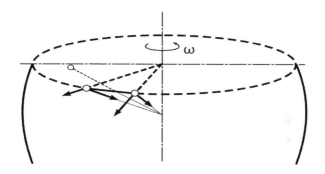

The horizontal bar of the equipment, shown in the figure above representing a cut-off view of Earth, was aluminum and the wire it was suspended by at the middle point was made of platinum with a diameter of about a thousandth of an inch. The two masses were both about an ounce and located on the ends of the horizontal bar. They were, however, of different materials, one made of copper and the other one was made of various different materials, some heavier, such as lead and some lighter such as wood.

Such an arrangement was obviously very sensitive to the torsional moment exerted on the string by the bar. Eötvös also devised a genial way of measuring the anticipated small torsional movement. He affixed a light source to the string and pointed the light beam onto a scale viewed by a telescope. The tiniest torsional movement would move the light on the scale visible by the telescope.

Eötvös carefully aligned the horizontal bar in east-west direction and calibrated his equipment to the center of the telescope scale. He then turned the bar around 180 degrees, so the ball that was on the east was on the west and vice versa. The result had not changed. He repeated with various balls on one side against a fixed copper ball. He found no difference. This is where the experimental physicist then became the theoretician.

His conclusion, by his own words was as follows: "In Budapest the centrifugal force changes the direction of the force of gravity by 356 arc seconds. If the gravity for the two balls of same mass but different material would be different by one part in a thousand, then the difference between the directions of the gravity forces would be 0.356 arc second, a third of an arc second. If the difference would be one part in twenty million, then the difference between the directions of the gravity forces must be larger than 356/20000000 or about 1/60000th."

He did not chose that number randomly, that was the resolution of his telescopic measurement. Since he did not notice any difference, he concluded that the gravitational force acting on the copper vs. the lead differs in less than one part in 20,000,000. This was a spectacular result, and Eötvös' experiment was accepted as the final proof for the equivalent effect of gravity on different materials.

To visualize how such an accuracy could be achieved one can use a laser pointer commonly used in presentations. Place the pointer on the desk in the back row of a room and align it straight to focus on the middle of the projection screen. Then move the pointer slightly, let's say by one degree. The light point on the screen will move, depending on the distance to the screen, by several feet. Let's say one degree moved the point to one side by 5 feet, or 60 inches. Imagine that on the screen there is a horizontal line scaled at every thousandth of an inch. That is not necessarily visible by the naked eye, but certainly by a telescope. Hence we just created the resolution to be able to measure 1/60000th of a degree.

In contrast, Galileo was about 2 in 100 accurate with his rolling ball experiments. Newton observed the period of pendulums of different mass but identical length and was about 1 in 1000 accurate. Later Eötvös increased his original accuracy to 1 in 500 million. He did this by improving his equipment, enclosed it to avoid interference from air and in his equilibrium calculations he even included the lift generated on the bodies by air. About two decades after his death a student of his reached an accuracy of 1 part in 2 billion. This number became 1 in 100 billion in the 1960s and gravity's independence from the material of the object became accepted as a proven fact.

There was an extremely important consequence of the above result. It was well known that the free fall motion of a body under the effect of the gravity force and the inertial motion of the body due to the centrifugal force both obey Newton's second law, the well known force = mass · acceleration. However, at that time it was not proven that the mass of the body under the acceleration of gravity, the gravitational mass, and the mass under any other force, the inertial mass are the same.

Eötvös' equipment should also have measured any difference that would have been due to the different effect of the gravitational and centrifugal forces. This was, however, not detected in the experiment, therefore the mass was behaving identically under both sources of forces. This is an experimental proof that the inertial mass is identical to the gravitational mass, there is only one mass. This became known as the equivalence principle and became a fundamental component of Einstein's gravity theory.

Eötvös was fascinated by mountains. He was an accomplished climber and even has a 2837 meters tall mountain peak named after him, Cima di Eötvös in the southern reaches of the Alps in Northern Italy, called Dolomites. It is highly likely that Eötvös was thinking about his science on his climbs.

Whether this was how it happened or not is not known, but he realized that mountains could result in a strong local change of the direction of gravity as well. He set out to modify his curvature variometer to be able to capture that. The new version was still built from two masses, a rod and a string. However, the two masses were now rearranged, one was still on the end of the bar but the other was hanging 2 feet below the bar. He called the new equipment horizontal variometer.

The idea of the device, now known as Eötvös pendulum, was that when there was an influence of a large mass nearby the device, for example a mountain, the gravitational direction would have a noticeable horizontal component change. As a consequence, the horizontal force acting on the two masses would be different and the horizontal bar would rotate in its plane.

The first successful measurements Eötvös executed with the equipment was to measure the influence of the mass of Mount Gellert across the Danube from his university laboratory on the Pest side of Budapest. In fact as a result of that he was able to measure the mass of the mountain, a first experiment of such kind ever.

He then redesigned his horizontal variometer to be portable and allow delicate experiments outside of a laboratory environment. He enclosed the strings into cylinders to avoid the influence of airflow, i.e. the wind in outside experiments. He insulated against heat absorption or loss and installed the ball of the horizontal bar on a thread to be able to adjust its position.

Eötvös decided to calibrate this portable version on the frozen surface of Lake Balaton one harsh winter, because that enabled him to eliminate the horizontal disturbances by any nearby mountains. The outcome of this activity led to a huge benefit to humankind.

He recognized that the gravitational direction also changed on the boundary of underground deposits of certain minerals. He proposed to use the technique for geological surveys and mineral explorations. He himself used his equipment to find the first oil deposits in Hungary by measuring the inequality of gravity between land and deposit masses. The device became known and used worldwide in oil exploration and made Eötvös a household name at least in science and the oil industry.

The fact that the local variation of gravity is related to the Earthly mass distribution is demonstrated on a grand scale in Canada. The gravity in the area of Hudson Bay has been found to be significantly less than the average gravity on other places of Earth at similar elevations. The anomaly is now attributed to the since melted Laurentide ice sheet that was at its time more than 2 miles thick and made a deep indentation on Earth's crust by pushing some of the material aside.

When the ice sheet finally melted, the indentation was filled in by water and the displaced material accounts for the missing mass and consequent lower gravity. The effect is not only noticeable by Eötvös' equipment; an experiment measuring the level of gravity by the motion of a pair satellites orbiting about 500 kilometers above the Earth and about 200 kilometers apart also proved it. When the satellites were above the Hudson Bay they moved farther away from Earth, due to the decreased gravity and the resulting difference with the balancing centrifugal force.

We now understand the concept behind the GRAIL mission: detecting Moon's interior composition by measuring gravity. There is, however, another mystery ahead of us: can gravity affect light?

7

Ether streams

Light is a topic that interested, influenced and intrigued humankind for millennia. The speed of light was also a very highly researched and contested subject. In fact one train of thought is that if the speed of light is finite, this in itself proves the presence of ether, Aristotle's fifth essence filling the space between celestial bodies.

After all, the photons, the hypothetical particles of light must be held back by something to a particular finite speed, otherwise they should travel with a faster, possibly infinite speed. Somewhat along the lines of the principle that Aristotle stated long time earlier as force = resistance · speed.

Galileo himself had already attempted to measure the speed of light with an experiment between two lamps equipped with shutters and placed a long distance from each other. The participants were indicating the detection of light from the other source by removing their own shutters, hence flashing a beam of light. The limited accuracy of time keeping devices of the time made the measurements rather crude, but still he arrived at the fundamental result of the speed of light's finiteness.

The ultimately accurate scientific measurement is

credited to Albert Michelson in the later part of the
19th century. He was a contemporary of Eötvös, born
four years later in 1852 and in fact not too far in Prus-
sia, in a village now part of Poland. His parents emi-
grated to the United States when he was 2 and he grew
up in California. As a foreign born person, he needed
a presidential exception from Ulysses Grant to attend
the US Naval Academy. He graduated, served at sea
and later returned as an instructor of physics.

Albert Michelson

Michelson was obsessed with the speed of light. He
executed his first experiment as an instructor at An-
napolis in front of students in 1877 and his measure-

ment yielded 299,864 kilometers per second. He continued refining his measurements during the rest of his life and ultimately achieved a value for the speed of light in vacuum as 299,940 kilometers per second that is now considered to be the precise value.

The speed of light is still a topic of some controversy. Some scientists consider it actually an immeasurable quantity. The reason is that one of the two components of the simple equation, distance and time whose ratio is the speed of light, is also established with the aid of light. Even if we have an atomic clock of extreme accuracy we still have a problem with the distance.

Consider measuring the time of light between two distant mountaintops, as Michelson did in one of his experiments in California between Mount Wilson and Mount Baldy, about 20 miles apart. Such distances are difficult to measure on the ground and usually done by using radar. But radio waves are part of the electromagnetic spectrum as we'll see it shortly and as such travel with speed of light. Therein lies the controversy.

Michelson became the first American scientist to win the Nobel Price in physics for his overall achievements in the optical precision instruments and investigations. One of those in fact became his claim to fame. He was intrigued by and set out to prove or disprove the existence of ether, the essence originally proposed by Aristotle some 2500 years earlier.

The topic was far from being forgotten; many scientists carried on the concept, in fact attributed it to be the cause of earthly gravity as well. Leonhard Euler, the great Swiss mathematician in 1760 hypothesized that the force drawing bodies together is due to ether losing density near the masses. Bernhard Riemann, the famous German mathematician, as late as in the 1850s argued that ether streams that we cannot see were carrying the bodies together.

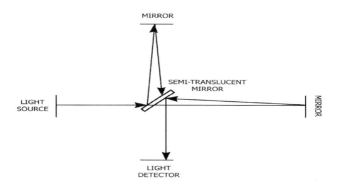

Michelson designed a device, shown in the figure above, he called the interferometer. The equipment was mounted on a rotating platform to allow changes in its orientation. In the center there was a semi-translucent mirror allowing half the monochromatic light originated on one side to transit and half to reflect perpendicularly. Both light paths ended in a fully reflective mirror and they were returned back toward the center. The light beam that was reflected first was now transiting uninterrupted and the light beam that was transited first was now reflected. They both ended up in the same light detector after having traversed the exact same path length.

The concept of the device was to measure the difference in the speed of light in the two directions upon arrival at the detector. Obviously that time was not measured by a clock, but by observing the interference pattern on the detector. If the light beams arrived at the same time, there would not be any interference noticeable. In the case of interference, the pattern distance would be indicative of the time difference and in consequence the velocity difference.

Michelson postulated that when the light is moving against the ether surrounding Earth it is slowed down and its speed is unaffected when moving in a perpendicular direction. If any difference were found that would prove the presence of ether. Earth's orbital velocity was estimated to be 30 kilometers per second, hence this was what they expected to find in their experiment if ether existed.

The experiments with Michelson's device were executed in 1887 in Cleveland at the Case School of Applied Science in Cleveland, Ohio, the place Michelson moved after leaving the Navy. His partner was Edward Morley, whose name is now forever associated with the experiment also. The experiments resulted in about 5-7.5 km/sec discrepancy between the speeds. While this was a nonzero value, they considered it small enough to be caused by the equipment and their measuring error. Then the scientific community jumped on the bandwagon and universally accepted that the experiment conclusively disproved the existence of ether.

There were many scientists questioning the accuracy of the equipment. Therefore the experiment was later repeated by another physicist, Dayton Miller on Mount Wilson in California, with a more refined equipment and better controlled environment. Miller's equipment was mounted on a mercury bed to avoid vibrations and the light beams went to a distance of 200 feet.

The Miller experiment measured about 8-10 km/sec speed difference, very close to the Michelson-Morley result but still nonzero. However, the decision about ether was already a foregone conclusion, so Miller's results were just considered to be the final validation. The fact of his more accurate measurements produced a value not smaller, but actually greater, did not raise any attention so much so that modern books now report that the experiment actually produced zero speed difference. That is of course factually incorrect.

There are still some who question the validity of this leap of logic and the ignorance of the measured nonzero speed. There is also the logical argument that not being able to detect ether does not mean it does not exist. Since these views have some justification, the topic of ether is alive and it will still be with us even in the last chapter.

The Michelson-Morley experiment, as it is now known worldwide, was later considered to be a foundation of Einstein's gravity theory that we will see next, even though Einstein said that his theory does not need ether.

8

Heavenly fields

When we are sitting in front of a fireplace and enjoying the warmth it generates, we are in fact sitting in a physical field, a heat field. A physical field's fundamental characteristic is that there is a certain quantity associated with points in space. In the case of the cozy space in front of our fireplace it is the temperature. This is a simple, scalar number; hence this is called a scalar field.

There are fields where a directed quantity (for example force) is associated with every point in space; such fields are called vector fields. The Newtonian gravity theory is described by a vector field. The acceleration of gravity at any point in space may be presented as a vector pointing toward the source of gravity and its value is related to the magnitude of the acceleration. This heavenly field could be visualized by imagining concentric circles around Earth where at each point of the circles an arrow would be pointing to the center of Earth.

The arrows representing the vectors would be the same size on a particular circle, but ever decreasing in length on the circles farther out starting from the surface of Earth. On the other hand, we have ever decreasing vector sizes on the concentric circles below the surface of Earth as well, in agreement with our earlier mental experiment of falling through Earth. The closer we get to the center of Earth, the more mass is above and around us, and the force of gravity will ultimately be cancelled.

Of course this heavenly field is not really that homogeneous since the gravity is not generated by only one object. As we saw in the case of the Lagrange points, there are places where the gravitational field value becomes zero and places were gravitational forces coalesce. Hence the Newtonian gravitational field of the heavens is in general non-homogenous.

Then there are even more complex fields whose associated quantity is a tensor, hence these are called tensor fields. A tensor is a square matrix containing various vector quantities. Einstein's gravity theory is represented by a tensor field.

Albert Einstein, arguably one of the most famous scientists of all time, was born in Ulm in Southern Germany in 1879 and had his elementary and middle school education in Munich. He moved to Switzerland in 1896 and enrolled at the university. He was a lackadaisical student, content with studying only topics that captured his interest. Fortunately for humankind, that included physics.

Upon graduation in 1900, he was employed by the Swiss patent office in Zurich. During the next two decades he produced his best scientific work resulting in the Nobel prize and a world-wide adoration of "the scientist who discovered the truth about our physical

Albert Einstein

world", quoted from the headlines of the times.

Einstein described a gravitational field that is not the result of the Newtonian force of gravity, but it is produced by a distortion of space due to the presence of mass. He also proposed that falling objects move along the lines of shortest distance, called geodesics, in this curved space. Geodesic lines are, for example, the routes taken by airplanes traveling long distances on the spherical Earth. They are the shortest distances over the curved (spherical) surface but not straight lines.

The crucial difference is in the fact that in Newton's gravitational field space is flat (Euclidean) and space in Einstein's theory is curved (non-Euclidean). Imagining the three dimensional space being flat vs. curved is a bit difficult and the words in the parenthesis open up the road to some deeper mathematical explorations, so we will just leave it at that.

Einstein was famous for his thought experiments, describing hypothetical what-if scenarios of physical phenomena. His theories were mainly based on those; he let others prove them experimentally, or he based his theories in some cases on others' experimental observations.

One such experimental observation was Eötvös' test and the resulting principle of equivalence. Specifically the fact that the balls of Eötvös' experiment were in equilibrium meant that the effect of the forces of the two different sources, centrifugal and gravitational, were identical. This became the fundamental underpinning of Einstein's gravity theory.

Einstein strengthened the statement in the form that is now known as the strong principle of equivalence. He declared that there is no difference between the inertial and gravitational motion of a body. He used a hypothetical experiment to illustrate this. He compared the phenomenon of a person tossing a ball from his hand to the ground on Earth, to the person doing that while traveling in a space ship with the same acceleration as that of gravity. The trajectory of the ball would be the same, he stated.

Einstein's gravitational theory is based on this principle, which is of course still only experimentally proven.

Einstein had to accept this as a postulate and built his theory on top of that. Without this his theory of gravitation falls apart. This possibility gives a potential attacking point to the theory that is still not fully endorsed by all scientists.

On the other hand, if the strong equivalence theory is correct, the consequence would be that the trajectory of any moving object is an intrinsic characteristic of the geometry of space. This is the essence of the geometric theory of gravity, as Einstein's gravitational theory is often called. Einstein's description of space being deformed by large masses made several unexplained celestial phenomena sensible.

Einstein's gravitational theory is represented by a compact equation containing two tensors. One of them, Einstein's geometry tensor, the solution of his equation, is a 4 by 4 matrix of numbers containing the information about the geometry of the curved space. It defines the geodesic trajectory of an object in the gravitational field governed by some mass or masses.

The effect of the governing mass or masses is described by another tensor that is also a 4 by 4 matrix of numbers. That is called the energy tensor and it represents the quantitative characteristics, in essence the strength, of the gravitational field.

We may visualize Einstein's curved space (for the moment in two dimensions) as a large flexible sheet, like the one the fire fighters (at least in movies) hold stretched out for falling bodies. If we imagine a large ball in the center of the sheet, we can envision that it

is going to deform it. The curvature of the deformation of the sheet will be larger the closer we get to the
ball producing it.

Rolling a smaller ball somewhere on the sheet, it
would obviously roll straight if the sheet were flat, as
shown by the straight line in the figure below. Due to
the indentation in the sheet, the ball will move into it
and its momentum will ultimately carry it out. From
the observer's point of view on the right hand side of
the figure, it will appear as the ball is coming from a
different direction.

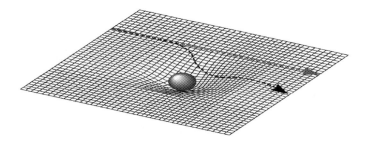

It is important to note that Newton's constant of
gravity G is still a part of Einstein's theory. This im-
plies that independent of the mathematical appear-
ance of the theory of gravitation, it is a universal constant. That is still only proven experimentally, but by
now to the accuracy of 1 part in 10^{11}, a number very
large even by astronomical standards.

The complexity difference between the two theories is notable by the differences between a Newtonian vector simply pointing toward the big mass (with some adjustments for multiple masses) and Einstein's geometry tensor with a 4 by 4 matrix of numbers (albeit only 10 of them distinct). The consequence is that the solution of Einstein's equation is much more difficult. In fact he himself only solved it approximately.

Einstein's approximate solution still led to a big success. Einstein conjectured that the behavior of Mercury is not exactly according to Newton's law of gravity because very close to the Sun the gravity field is too strong for Newton's equation to be accurate. He applied his solution for this scenario and found Newton's missing 43 arc seconds. Einstein was so happy about this, he considered it to be the vindication of his revolutionary approach.

It is worthy of pointing out that Newton's theory can very easily be fixed by using the power of 2.00000016 instead of the square (power of 2) used with the distance in Newton's formula. That minute correction would describe the Mercury precession just as accurately as Einstein's. If this is not considered to be scientifically prudent, we must also mention that Einstein did something similar.

Einstein's equation produced the undesirable (at least to Einstein) outcome of the universe contracting under gravity. The prevailing cosmological model at that time was a stationary universe, hence that contradicted Einstein's equation. To counteract this, Einstein introduced a factor into his equation, called a cosmological constant. Clearly this was also a heuristic adjustment similar to the one above with Newton's formula.

Then he was confronted by the expanding universe hypothesis of Edwin Hubble in the 1930's, another topic of prevailing scientific belief that is still occasionally contested. Hubble's observations made Einstein's factor unnecessary and he was forced to recant it. Einstein was so embarrassed by it, he called it the greatest blunder of his life.

For weak gravitational fields and low speeds compared to the speed of light, Einstein's more complex mathematical system reduces to Newton's. This is the reason why Newton's theory is still used in most everyday circumstances. Einstein's more complex theory, however, provides mathematical consequences that lead to very interesting gravitational phenomena.

For example, let us return to our flexible sheet figure and replace the ball with a light beam. The consequence of Einstein's theory is that light should also be bent around a large mass. Einstein's curved field forces the path of light passing close to the dominating mass of the space to follow its curvature. The closer the beam is getting to the mass, the more aggressive the bending will become.

This phenomenon is called gravitational lensing. As implausible it sounds, it was experimentally proven as we shall see next.

9

Bending waves

Not easily, but by the end of the 19th century we ultimately arrived at the understanding of light's wave nature as being a kind of electromagnetic radiation. It was a road marked with the ongoing scientific animosity between Hooke and Newton, whose conflict regarding the credit for the discovery of the inverse proportionality of the force of gravity to the distance squared was mentioned in an earlier chapter.

Regarding light they were on the opposite sides of the spectrum. Hooke published a work in 1660 in which he argued that light must have a wave nature. Newton on the other hand was an ardent proponent of the conventional belief about the particle nature and wrote about it in 1675 in a work titled "Hypothesis of light". In this work Newton proposed that light was comprised of particles (the hypothetical photons) of matter that traveled in straight lines.

Despite being wrong, due to Newton's higher standing as scientific authority, his view was accepted and ruled for more than another century and a half. There were, however, some problems with the theory and the belief started to erode in the first half of the 1800s. While the particle theory was able to explain the simple reflection of light, it was in trouble with the refraction which was a well known phenomenon experienced in daily life.

Wave theory also had arguments against it. If light was a wave, it would need a medium to travel in, so went the conventional wisdom. In the case of sound waves, a phenomenon well understood by that time, the pressure wave generated by the sound source traveled in the air. What was the medium carrying the light waves between celestial objects? Aristotle's ether came to help again and was proposed to be the medium carrying the light waves.

But there was another phenomenon of light that could not be explained by the particle nature: the polarization of light. This started the avalanche that ultimately brought down the particle theory, but it took a collection of scientists to be able to do it. Incidentally the main players were all British; it appears that only they had the conviction to attack their famous countryman's sacrosanct legacy.

It started with Thomas Young in 1800, who executed various diffraction experiments with horizontal and vertical slits placed in the direction of light. Young demonstrated the interference phenomenon utilized by Michelson later in his equipment. He already hypothesized that the different colors are simply light with different wavelengths, a topic soon introduced.

About 45 years later Michael Faraday discovered that the plane of polarization of light rotates when traveling through a magnetic field. That was a huge clue pointing toward light also being an electromagnetic phenomenon. He also conjectured that light was a high frequency (another concept soon explained) wave that could propagate even in vacuum without the need for a medium.

The final proof was produced by James Maxwell who in 1873 published a work in which he provided a clear mathematical model, now known as Maxwell's equations, that described the behavior of all electromagnetic phenomena, including light. Even the speed of light issue was put to rest by Maxwell whose equations analytically confirmed the speed of light measured by Michelson.

Let us now investigate this important aspect of light, its wave nature. Waves are of course all around us, in the oceans and in human waves at stadiums. They are are characterized by two main quantities that were mentioned above, the wavelength and the frequency. Wavelength is the distance between two identical positions considering also the direction of motion, otherwise known as phases. Frequency of the wave is a closely related fundamental characteristic; that is the number of repeated waves in a certain time interval.

For example the number of waves in a second is called Hz (for German scientist Rudolf Hertz). The frequency and the wave length are inversely proportional, the longer the wave length the lower the frequency and vice versa. They are tied together: their product is the speed of the wave.

The type of the medium the wave is traveling in will strongly influence the speed of the wave. It is well known that the speed of light is different in water as opposed to air and results in various diffraction scenarios. Light is a certain section of the larger class of electromagnetic waves as shown in the next table.

Range	Wave length meters	Frequency Hertz
Radio wave	10^4	10^4
Microwave	10^{-1}	10^9
Infrared light	10^{-4}	10^{12}
Visible light	10^{-6}	10^{14}
Ultraviolet light	10^{-8}	10^{16}
X-ray	10^{-10}	10^{18}
Gamma ray	10^{-12}	10^{20}

All of these waves travel with the speed of light in vacuum. As we learned earlier the product of the wave length and the frequency produces the speed of the wave. The product for each wave type is approximately 10^8 meters per second, that is 10^5 kilometers per seconds, roughly the value of Michelson's experiment.

The mystery of how would light be affected by gravity was easy to resolve when light was thought to be of particle nature during Newton's time. It was a natural consequence of Newton's gravitational theory that light particles would be affected by large masses, like the Sun. This is somewhat different in view of light's wave nature.

The light emitted by a star located beyond the Sun

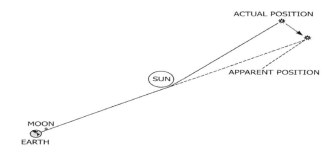

would appear to come from a different origin than its location. Such an observation could be made at the time when the Sun is obscured by Moon during an eclipse. An actual experiment, first executed by the British scientist Arthur Eddington in the first part of the last century, is shown in the figure above.

Eddington was born in 1882 in England and after his studies at the predecessor of the University of Manchester he earned a scholarship to Cambridge's Trinity college, one of the most prestigious schools of England. He graduated in 1905 and later became the Royal Astronomer at the Cambridge Observatory and even the secretary of the Royal Astronomical Society. In this latter capacity he conducted the experiments that made his name forever associated with the proof of gravity's effect on light.

Eddington prepared by taking night photographs of the Hyades constellation to obtain reference pictures. He also devised a comparison mechanism of the reference plates with the future photographs to be made on the site of observations. Eclipses of course are rather predictable and one was coming on May 29 in 1919 during which the Sun was going to be in the direction of a set of very bright stars in the Hyades constellation.

Arthur Eddington

Eddington led a team to the island of Principe off the African coast and his team managed to make 16 photographs of varying exposures. Some of those were developed on site and they were already very promising in the comparison mechanism.

The ultimate results were not obtained until they returned to England and all the photographic plates (the technology at the time) were developed. But then the results were conclusive. The comparison of the plates and the reference plates showed the displacement of the stars in the expected direction from their actual position.

This was not just a qualitative experimental proof. Eddington's actual measurements of the amount of deflection due to the gravity of Sun was quantitatively the same as Einstein's theory of gravitation predicted. Eddington's experiment put Einstein's theory onto the front pages; in reality Eddington made Einstein famous in everyday circles.

Eddington received the highest scientific honor of England, the Royal Medal of the Royal Society in 1928 and ultimately had an honor named after him as the Eddington Medal of the Royal Society. He also received knighthood in 1930. Ironically Eddington, as Einstein, also spent many years of his life trying to find a theory which would have unified electromagnetism and gravitation. Neither of them were successful at accomplishing it.

Since Eddington proved the fact that gravity affects light in the transversal direction, it appears quite plausible that light waves are also affected by gravity in the longitudinal direction. In order to investigate this, we make a side trip to the sound waves already mentioned earlier.

The Austrian scientist Christian Doppler described an effect that we all also noticed. When a train approaches a station while blowing its whistle, the sound gets higher and higher pitched on approach, and deepens on departure. The phenomenon may be understood as the distance between the wave peaks (the wave length) is decreasing and the number of waves in a certain time period (the frequency) is increasing. Increased frequency means a higher pitch of the sound, just like the real life observation. This phenomenon is now called the Doppler shift.

Since sound and light are both waves, the same phenomenon is also observable with light. Assume that a light source is moving away from us, similarly to the train leaving the station, the light's frequency is getting lower just like the train's whistle is getting deeper.

Let us focus on the visible light range in the table of electromagnetic waves, the range of 10^{-6} meters. This may be further subdivided in the the range from 400 nanometers (10^{-9} meter) to 700 nanometers. The shorter wavelength hence higher frequency is the violet color that we observe. Then we proceed to blue, green, yellow until we reach red at 700 nanometers. This is the lower frequency end as shown in the table.

Range	Wave length nanometers	Frequency Megahertz
Red light	700	$1.43 \cdot c$
Yellow light	600	$1.67 \cdot c$
Green light	550	$1.82 \cdot c$
Blue light	450	$2.22 \cdot c$
Violet light	400	$2.50 \cdot c$

Based on the table where c denotes the numerical value of the speed of light, increasing frequency of light would move it toward the blue side of the spectrum and decreasing frequency of light would move toward the red side. This tendency is called the redshift.

This phenomenon is that on which Hubble's expanding universe theory relies when states that the redshift of light arriving from distant stars indicates that they are moving away from us. There is, however, another cause for redshift and that is by gravitation. This is a consequence of Einstein's theory implying that light waves leaving a very heavy and therefore strong gravity star are slowed down by the gravity field. This would result in the wavelength being shorter as shown in the following figure.

The frequency on the right hand side, the origin of the wave, is higher than the frequency of light in the interstellar space. When the light wave is farther and farther away from its originator star the gravitational force of it is ever decreasing and the light's wavelength is increasing.

We can reconcile this with the Doppler phenomenon. Let us imagine us being the observers at the left end of the figure. The wave pattern from that side appears to be an ever increasing wave length, or ever decreasing frequency. The increasing wave length of the light is a shift toward red, hence the name redshift.

This phenomenon has been verified by astronomical measurements of the spectrum of the Sirius system. This is a special dual system of a large (A) and a smaller (B) star. Sirius A has been observed since antiquity and the start of the ancient Egyptian calendar was set according to its rising. The dual nature enabled the redshift comparison, since they move around each other in synchrony. Since the larger star demonstrated a stronger redshift than the smaller, it indicated that the cause was gravitation and not expansion.

Some scientists take gravitational redshift to be the explanation of the redshift of all stars exhibiting the phenomenon. This puts Einstein's theory in collision course with the expanding universe hypothesis and it is strongly against the grain of the current scientific mainstream. This is not the only topic where gravity's mysterious behavior and effects seemingly contradict the ruling scientific consensus as we will see it later.

Despite the enthusiastic public acceptance of Einstein's theory having been "proved by Eddington" as the newspaper headlines said, the theory was not fully endorsed by the scientific community until the 1960's. By then there were more and more astrophysical observations supporting several more consequences of Einstein's theory.

10

Black holes

Black holes are singularities humorously described as the result of God trying to divide the universe by zero. The German scientist Karl Schwarzschild proved the possibility of their existence as a consequence of Einstein's equation in 1916.

Schwarzschild was a child prodigy already publishing papers about celestial mechanics at the age of 16. He was born in Frankfurt, Germany in 1873 and conducted studies in Strasbourg and Munich, earning a doctorate at the age of 23. After several years of heading an observatory in Vienna, he was appointed to be the director of the observatory at the famous university of Göttingen, home of such luminaries of science as Gauss.

After a decade and a half of fruitful scientific research, the outbreak of World War I changed his life. He joined the army and continued doing research work while on the Russian front. Incidentally that included finding an exact solution to Einstein's equation. Upon his return he sent his solution to Einstein and soon after he died as the result of a disease he developed on the front. He was honored by an asteroid named after him: 837 Schwarzschild.

Einstein's geometry tensor was a symmetric matrix

of 4 rows and columns. The four dimensions were the 3 spatial directions and time. The symmetry results in 10 unique terms, four on the diagonal and 6 above, these are the 10 solution parameters of Einstein's equation. Hence Einstein's tensor equation may also be described in 10 nonlinear, partial differential equations, the same types Laplace used and that are somewhat beyond our mathematical level here.

These are called Einstein's field equations and they are still extremely difficult to solve exactly. The 10 solutions of these equations (our memories from middle school math indicate that the number of solutions for a system of equations is equal to the number of equations) specify the ten distinct components of the geometry tensor and describe the geometry of the curved space.

Schwarzschild's exact solution was assuming a spherically symmetric but not rotating mass generating the gravity field. His solution produced a special term which is now called the Schwarzschild radius. The formula of the term depends on Newton's gravity constant, the governing mass and the speed of light.

The meaning of this radius is that if an object's geometrical radius is smaller than its Schwarzschild radius, it is a very high concentration of gravity, the aforementioned black hole. The perimeter of the circle with this radius is the so-called event horizon of the object. That is the neighborhood of the black hole from beyond which there is no return. Neither for mass, nor for light, hence the name.

Einstein himself was not too pleased about this possibility, after all, his approximate solution did not yield this scenario. Nevertheless, the presence of such objects is now widely accepted as being massive stars that collapsed under their own gravity. It is believed that the center of our own Milky Way galaxy is a giant black hole.

The black hole at the center of our galaxy is sometimes called a galactic glutton as it seems to have an insatiable appetite for material. The sizes of black holes are usually classified by their density. That is computed as the amount of mass in the black hole divided by its Schwarzschild radius. The amount of mass in the black hole of Milky Way is estimated to be 2 million times the mass of the Sun. Its density is assumed to be larger than the $1.8 \cdot 10^8$ mass unit per volume boundary, classifying it as a super massive black hole.

The event horizon for the black hole at the center of our galaxy is about 36 light hours, or close to 40 billion kilometers, a very formidable distance even in galactic terms. Popular science films depict the fate of an object going through the event horizon as being stretched out like a long spaghetti. It appears that this scenario is about to happen.

Recent astronomical observations show a giant gas cloud speeding toward the center of our galaxy with about 8 million kilometers per hour velocity. The close encounter will occur in 2013 at which time the event horizon hypothesis could be viewed in a very large scale.

There are many black holes around us. One super massive black hole named NGC 3842 (New General Catalogue of nebulae and clusters) is located about 320 million light-years from Earth in a cluster of galaxies in the Leo constellation. It is about 9.87 billion Sun masses. Then there is another one in the Coma constellation, subject of some observations in our last chapter. This is about 336 million light-years away with almost 20 billion Sun masses, and it is the present record holder.

Black holes are not only a dead end of collapsed celestial objects, in fact they are supposed to give us a glimpse of the formation of galaxies as well. Some scientists suspect that the center of every galaxy has a black hole and they are somehow related. The research is now executed with orbital observatories, such as the Hubble space telescope.

To put the concept of the Schwarzschild radius into human perspective, for the Sun it is about 3 kilometers and we know that the Sun is much bigger than that. Similarly, the radius for Earth is only about 9 millimeters and we know that its geometrical radius is about 6,400 kilometers. We are safe from Earth or Sun collapsing into a black hole soon.

The concept of the event horizon and Schwarzschild's radius brings another apparent contradiction with prevailing scientific beliefs. The big bang hypothesis proposes that there was an extremely small, but very dense kernel of primordial material that exploded and resulted in the material in the universe. The theory proposes that the bang happened in a time frame of only about 10^{-36} seconds and during that time an expansion of volume by a ratio of 10^{78} occurred.

These numbers imply that the universe at the beginning was much smaller than its event horizon and therein lies the contradiction. Einstein's theory of gravitation via its mathematical consequence of the black hole in Schwarzschild's solution proposes that nothing can escape from the region inside of the event horizon and that seems to fully contradict the big bang theory. Since the current size of the universe is of course much larger than its event horizon, there must have been a process to get material out of the original event horizon.

One way out of this contradiction is by saying that Einstein's gravity theory simply does not apply in these circumstances. The primordial environment was an extremely hot (perhaps millions of degrees of Celsius) soup of elementary components of material, whose nuclear interactions of extremely high energy overrode any other physical considerations. If that sounds like an unsupported argument, the next alternative is even more unproven.

This, rather farfetched explanation is the possibility of white holes. A white hole, as the name indicates, would be the opposite of a black hole. Material could not enter the white hole, but matter including light could leave it. The concept hinges heavily on the time component of the four dimensional space used in Einstein's theory and sort of proposes a temporal progression.

In this theory, the white holes are the past and black holes are the future. The material of a past white hole could explain the big bang hypothesis, and since that happened many billions of years ago, it fits the past history scenario. The fact that black holes are inevitable in the future will be soon validated by another theoretical consequence of Einstein's gravitational theory in the next chapter.

The white hole-black hole scenario is even more attractive in connection with the theories proposing the cyclically expanding and contracting nature of the universe: the big bang being followed by a big crunch. This in itself leads to a highly speculatory side road, so we'll abandon the topic.

The limitation of Schwarzschild's solution was that the spherical object was not rotating. That is, however, an everyday occurrence of celestial objects, hence attempts were made to find a solution for such case as well. Rotational black holes would be the result of spinning stars collapsing under their own gravitational field.

An exact solution to Einstein's equation for such scenario was produced in 1963 by Roy Kerr, a New Zealand mathematician. He extended Schwarzschild's model by allowing the rotation of the mass. This solution is considered to be the most important solution of the equation since it is the best fitting to our universal observations.

Roy Kerr, born in New Zealand in 1934, was also

Karl Schwarzschild   Roy Kerr

a mathematical prodigy like Schwarzschild. He completed his studies at the University of New Zealand as a teenager and had to wait for his formal graduation papers until he reached 20 because of regulations. He earned his doctorate at Cambridge University in 1959 with a dissertation already focused on Einstein's equations and before he turned thirty he presented his solution.

Kerr also received his share of medals for his outstanding work in theoretical physics, and ultimately the Companionship of the New Zealand Order of Merit in 2006 "for his services to astrophysics".

Kerr's solution also produced a black hole, but with two singularities. The inner singularity was a spherical event horizon, similar to Schwarzschild's. The outer singularity was a surface of a flattened sphere, not unlike our Earth. Between the two singular surfaces the black hole would be still rotating with a gradually diminishing speed. The rate of change of this rotation speed depends on the angular momentum and mass of the originating star. It is now believed that the black hole at the center of our galaxy is also rotating.

Rotating black holes may be created when a rotating object's angular momentum reaches a critical minimum that is a function of its mass, Newton's gravity constant, and the speed of light. This is along the line with the reasoning we followed earlier in connection with Moon's motion. The angular momentum held Moon speeding along on its orbit balanced by the gravitational force. As we mentioned there, if Moon would loose its angular momentum, it would fall to Earth.

Our Sun, formed about 4.57 billion years ago, is an interesting example since it also has a variable rotation. At its poles it rotates in 35 days and at its equator in 25 days. Its estimated mass is about $2 \cdot 10^{30}$ kilograms and accounts for almost 99.9 % of the mass in the solar system. The diameter of the Sun is 1,392,000 kilometers and its angular momentum is approximately 5 times the critical minimum.

It appears that we are on safe grounds with Sun and it will not become a rotational black hole in the near future. Mysterious gravitational phenomena, however, do not end at the black holes, objects orbiting large masses encounter another remarkable phenomenon.

11

Falling skies

Newton established the fact that Moon is falling while kept in balance by its rotational energy. But how long will that balance remain? Is there any reason why a rotating planetary object would lose some of its rotational energy? The answer is yes and the result is a phenomenon called orbital decay.

Moon and Earth are in a very interesting relationship called tidal locking. That is the reason why we always see the same face of Moon. The rotational frequency of Moon around its own axis, and its orbital frequency around Earth are the same. This is a frequent relationship between moons and their celestial hosts in the solar system, and a preferred arraignment of man-made satellites.

The orbital decay phenomenon in the case of man-made satellites that are affected by Earth's atmosphere is obvious. The resistance of the atmosphere, called atmospheric drag is going to have a spiraling effect: the craft is going to lower its altitude, the resistance will be higher at lower altitude, and the process is repeated until the craft heats up due to resistance and burns before hitting the ground.

In the case of celestial objects, the decaying of orbits is manifested by the gradual deformation of the elliptic orbits toward circular shapes and a reduction in the radius. The total rotational energy, or angular momentum of the system (assuming a two body system as discussed earlier in connection with the Lagrange points) will decrease by the amount of the energy carried away in the form of gravitational waves.

These waves are predicted by Einstein's theory of gravity as follows. When an object moves the geometry of the space changes its curvature and the change propagates with the speed of light. This is the gravitational wave and it carries energy (in case of rotating objects the angular momentum) away from the object.

In order to visualize gravitational waves, let us revisit the flexible sheet with the big ball in the middle from the earlier chapter. That is sort of a stationary scenario, the ball (mass) and the distortion of space are at equilibrium. Let us now consider a smaller ball rotating around the big one at the perimeter of the indentation, as a planet rotates around its sun. The smaller object also creates its smaller indentation, however, this is not stationary. It is moving with the object along the orbit.

The dynamically moving indentation creates small ripples in the fabric of the sheet, as well as in the geometry of space. These ripples are the gravitational waves. The rotating object will lose some of its energy due to this combined effort of creating its indentation and ripples, resulting in the decaying orbit. In the case of the flexible sheet example, the smaller ball rotating around will be slowly spiraling toward the bigger one resulting in the phenomenon called in-spiraling.

The orbits of the planets are very close to being circles. For example, the eccentricity, the deviation from a circle of Earth's orbit around the Sun is 0.0167. This measure is zero for a perfect circle and one for an ellipse, clearly Earth's orbit is for any practical purpose is circular.

Most of the orbits cannot be described as the result of the gravitational interaction between only two objects. Even the motion of the Moon cannot be accurately described by only considering Earth's effect. Once the Sun is also included, the scenario presents a very hard so-called 3-body problem that can only be solved approximately in most cases.

Assuming a two-body relation, however, the rate of the orbit reduction of planetary objects may be computed as a function of the distance between the objects, their masses, the speed of light and Newton's gravitational constant G. As a consequence of that, the actual time the object will stay in orbit, called the orbital lifetime may also be computed with the same parameters, give or take some constants.

The orbital decay computation for the Sun-Earth pair results in approximately 10^{-20} meters per second or about 10^{-13} meters orbital decay per year. This is an extremely small number, about two orders of magnitude less than the diameter of a hydrogen atom.

The orbital lifetime of Earth is in the order of 10^{23} years that is 10^{13} times longer than the current estimated age of the universe of 14.3 billion years. Hence we do not have to be concerned about Earth falling into Sun soon.

Orbital decay was observed outside of the solar system by two US astronomers, Russell Hulse and Joseph Taylor, working at the Arecibo observatory in Puerto Rico found a binary system whose one member was a pulsar. This was a recent discovery in 1974, barely a quarter of century ago, definitely miniscule time considering our book's time horizon that began with Aristotle. Their discovery brought a quick Nobel prize for them less than 20 years after their discovery in 1993.

Joseph Taylor   **Russell Hulse**

Pulsars emit periodic radio signals that are detectable on Earth. The stars are massive, about 40 % larger than our Sun, but they orbit each other extremely closely, in a distance of just over half the distance from Earth to Sun. Hulse and Taylor predicted that their close proximity and large mass would generate a large energy loss that could be detected.

Their long running observation demonstrated that they are indeed on a spiral path toward each other as a result of the orbital decay. That was proven by the Doppler shift in the radio waves and the measured results produced a good match with the prediction from Einstein's gravity theory.

Gravitational waves are assumed to have the same characteristics as simple waves introduced in an earlier chapter: wave length and frequency. Their product is the speed of the wave, specifically speed of light. This explains the instantaneous effect of gravity as detected by us and other physical phenomena.

There are ongoing, but yet unsuccessful attempts at measuring gravitational waves in laboratory environment with experiments. The experiments are executed with devices that are sort of a marriage of the Eötvös variometer and the Michelson-Morley instrument.

The concept of such experiments is to differentiate between the gravity waves induced motions of masses that are separated by large distances on arms arranged perpendicularly to each other. The gravitational waves would stretch one arm of the device by moving the masses relative to each other aligned in the direction of the wave and not affect the other.

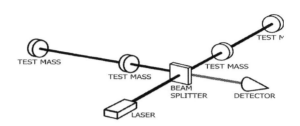

The experimental devices, as shown in the figure above, usually contain two masses connected by arms that are several kilometers long and oriented in east-west and north-south directions. The arms are actually vacuum tubes and a laser beam is traveling in their interior. Highly reflective mirrors are installed at the end of the arms to reflect up to 99.9 % of the light. The recombined laser beams are to be detected in a laser interferometer. The goal is to detect a stretching of the arms as small as 10^{-18} meters.

The phenomenon of orbital decay was due to energy loss by gravitational waves as a consequence of Einstein's theory of gravitation but, despite several practical realizations of the above concept in the US, China and Europe, gravitational waves have not been detected yet. On the other hand, there are other phenomena that are already detected, but not explained by Einstein's theory.

12

Celestial slingshots

The technique we call celestial slingshot is based on temporarily entering into the gravitational field of a planet and using it to increase the velocity of an object. This increase in velocity is with respect to the Sun, assuming we are considering objects in our solar system. It does not change the velocity of the object relative to the planet whose gravitational field is used for the assistance.

Let us imagine the scenario of an interplanetary object approaching a planet with a velocity v while the planet is moving in the direction of the object with a velocity of u. Both of these velocities are with respect to the solar system. If the object achieves a close enough path to the planet to be captured by its gravitational field, it will temporarily enjoy its gravitational pull. Due to this pull, the object is going to attain a parabolic or hyperbolic orbit around the planet. Because the velocity of the object is larger than the escape velocity of the orbit, the object will leave the planet's gravitational field in the opposite direction.

During the approach the relative speed of the object and the planet was $u + v$ as they were moving toward each other. Upon leaving the planet's gravitational hold, the object's relative speed to the planet is still $u + v$. However, since they are now moving in the same direction, the object's speed relative to the solar system is $2 \cdot u + v$. In summary, the object gained an additional $2 \cdot u$ speed from this planetary encounter. In practical scenarios the incoming and outgoing paths of the object are not parallel, hence the boost obtained is less than in our hypothetical example, but still significant.

The concept of gravitational slingshot was first proposed by a Ukrainian scientist, Yuri Kondratyuk. Kondratyuk was born in 1897 in the city of Poltava in nowadays Ukraine, then part of Russia. He received engineering education at the predecessor of the Saint Petersburg Technical University and in his case his father was the one who directed his interest toward science. He was especially interested in interplanetary travel and his idea of a two stage vehicle to travel to the Moon was actually used in the Apollo program we discussed earlier.

Kondratyuk was born under a different name, but because of his involvement in the Russian revolution on the Tsarist side he was forced to change his name to avoid persecution. He went into a self-imposed exile to Novosibirsk in Siberia in hope of escaping the attention of the Communist regime. While there he wrote a book titled "The conquest of interplanetary space" that contained his original ideas, such as the subject of this chapter. Kondratyuk was later drafted into the Soviet army in 1941 and died on the western front less than a year later, adding to the sad loss of Schwarzschild of an earlier chapter.

Yuri Vasilievich Kondratyuk

He was later honored by naming a crater after him on the Moon. There is also an asteroid 3084 Kondratyuk named after him. His most important posthumous recognition was likely the visit by Neil Armstrong who went to his Novosibirsk house, gathered a handful of earth from the garden and took it back to the US.

Real life applications of the technique are the probes we sent out to various destinations in the past. Sending probes toward the inner planets, located between Earth and the Sun, is somewhat easier since the craft travels toward the Sun and accelerates due to the Sun's gravitational pull. On the converse, sending spacecraft toward the outer planets is more difficult because the craft must move against the gravity of the Sun, hence in this case the gravitational sling is useful.

The first use of this technique was by the Mariner 10 spacecraft launched in 1974. The craft used the gravitational field of Venus to accelerate. It is now about 17 billion kilometers from our Sun. It is in the outskirts of our solar system and is in transition to interstellar space.

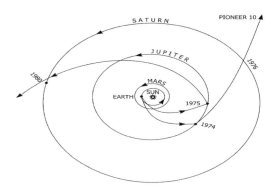

Another notable application of the technique was by the Pioneer spacecraft, 10 and 11, whose paths are shown in the above figure. They were launched in the early 1970's to explore the outer reaches of the solar system and ultimately leave the solar system in opposite directions. They were sending electromagnetic signals dutifully back for more than the designed 20 years of duration of their lifetime. In 1998 it became obvious to the controllers that both craft went astray beyond the orbit of Pluto, the since deposed planet now called planetoid.

The sudden change in the curves depicting the path of the spacecraft at the orbit of Jupiters demonstrates the gravitational slingshot. In fact this technique itself also produces a gravitational mystery, known as the flyby anomaly. The anomaly presents itself in an unexplained velocity change during such maneuvers. The anomaly was noted with several probes in their Earth flyby maneuvers. The first was the Galileo spacecraft in 1990, the NEAR (Near Earth Asteroid Rendezvous) spacecraft in 1998 and the Rosetta spacecraft in 2005. There are several possible causes of these anomalies.

The anomaly could certainly be due to an incorrect prediction of gravitational force by Einstein's gravitation theory. It is conjectured, however, that this theory augmented with a consideration of rotation of Earth, the solar system and even the universe would properly predict the behavior. The other, more controversial proposal is that Earth is surrounded by a halo (an extended spherical region) of dark matter. There is no consensus yet, so we will leave it at that, but surely another mysterious discrepancy between the theory and observation is lurking here.

The controllers of Pioneer 10 were able to establish that the craft was off course by approximately 400,000 kilometers. That is about the Earth-Moon distance, so if it was aimed to reach Earth from those outer regions, it would have arrived at the Moon instead. That is of course, if the unexplained cause works similarly on a return trip. Since we now see very far, there is no unexplained planet (another Vulcan), or large object that would account for the discrepancy. Einstein's theory is at a loss in explaining the discrepancy.

The discrepancy is actually due to a deceleration of both craft. This was realized by the blue-shift of their signals, the opposite of the red-shift we discussed at length earlier. The deceleration could point to internal causes in the spacecraft themselves and the other possibility would be external causes.

One of the internal causes considered by the NASA engineers was thermal radiation from the onboard electronics. It was hypothesized that some of this radiation could have been reflected from the antenna dish of the spacecraft and acted like a solar sail. Other internal causes could have been the course adjustments by the rocket thrusters of the craft, but ultimately all internal causes were eliminated.

That left the possible external causes. One possible external cause could be if Sun's gravity at that distance was bigger than predicted by Einstein's gravity theory. While the theory was vindicated by inner solar system observations, its validity was never really demonstrated with outer solar system objects. This would potentially mean an end of Einstein's theory in a similar manner to how Newton's was deposed due to the observation of Mercury's orbit.

Another external explanation would be the presence of something in space slowing down the craft. This plausible, yet unproven idea leads to our next mystery.

13

Dark matters

It is now well known that there is some discrepancy between the masses of certain galaxies computed from their luminosities and the mass required to produce the gravity to balance their rotational behavior. Specifically, the latter is larger, leading to the conclusion that there is matter that we do not see, hence called the dark matter.

The dark matter name is credited to the colorful Swiss astronomer Fritz Zwicky. He graduated from the Swiss Federal Institute of Technology in Zurich in the 1920s and later moved to the California Institute of Technology where he worked for the rest of his life.

He was somewhat of an oddball and a prankster at the institute. He was not much liked by students or other faculty members. One of the anecdotes about him says that when cloudy weather disrupted his work, he instructed his assistant to shoot a gun in the air to chase away the clouds. This was at the famous Mount Wilson observatory, incidentally the site of Michelson's and Miller's activity, where Zwicky spent most of his time observing the skies.

Despite these musings and personality quirks, he was considered to be a brilliant astrophysicist with an important scientific legacy. In the 1930s he was observing the speed of the Coma super-cluster (NGC 4889), the home of the current record holder black hole. Zwicky noticed that it was about 160 times bigger than should have been based on the visible mass in the cluster and as such contradicted Einstein's gravitational theory.

Fritz Zwicky

In the case of clusters, the mass generating the gravitational force is the combined mass of all the visible galaxies. According to Zwicky's calculations, the clusters should have dispersed a long time ago, losing the outlier galaxies first due to the insufficient gravity arising from the visible mass. That would have lowered the amount of mass remaining, further lessen the gravitational force resulting in losing more galaxies. This process would have continued until the death of the cluster.

Since the cluster was stable, Zwicky concluded that there must be extra mass in the system that is not visible and the dark matter phrase was born. This hypothesis slowly gained ground and by now is universally accepted despite the lack of any direct proof. In fact it is now assumed that dark matter could take up as much as 95 percent of the matter in the universe.

Speculations surrounding dark matter are swirling. Dark matter may also account for the discrepancy of Newton's theory with respect to Mercury's precession that would render the Newtonian gravitational theory accurate again. And dark matter could be the missing piece of Einsteins gravitational theory to account for the Pioneer anomaly.

There are some who consider the introduction of dark matter just an excuse to avoid a radical change of the existing gravity theories. This is certainly easier than reinventing gravity as the title of one of the references proposes. It appears that three possibilities exist: modifying Newton's or Einstein's theory, or inventing a brand new one.

Modifying the Newtonian gravitational theory is proposed by applying an adjustment to it in close proximity or far away from the origin of the gravitation. It is assumed that an adjustment in the gravitational acceleration at the outskirts of the solar system on the order of 10^{-10} meters per second squared would explain the observed anomaly of the Pioneer craft. This is not proven yet, but the proposal is not being completely discarded off hand by scientists.

Then there is another school of thought proposing to rework Einsteins theory of gravitation by generalizing the symmetric geometry tensor into unsymmetric. An unsymmetric geometry tensor would yield 16 parameters as opposed to the ten in the current form, albeit the meaning of some of the new parameters is not clear. The method brings serious mathematical difficulties in validating it against observations.

A brand new theory, if one were to attempt creating it, should involve the dark matter in one form or other. Interestingly, it appears that dark matter and ether, originally proposed by Aristotle 2500 years ago and described in our first chapter, are somehow related. Both seem to have the role to fill out the space invisibly.

As long as there is no direct proof for dark matter apart from the above indirect argument, one might rightfully entertain the possibility that Zwicky's dark matter is the equivalent of the ether of Aristotle. It seems like we covered the story from A to Z, certainly we arrived from Aristotle to Zwicky.

But there is another, rather vexing mystery of why one cannot find any relationship between gravity and electromagnetism. They are so much alike and so different at the same time. Both of their reach extends to infinity and their distance behavior is inversely proportional to the square of the distance. They both travel with the speed of light, although the speed of the propagation of the gravity field may only be the side effect of the Einstein's gravity theory. Just like light travels in the waves of the fictitious photons, gravity's effect could be carried by a hypothetical graviton.

This is where the main similarities end. Their relative strength is vastly different. The electromagnetic forces are about 35 orders of magnitude stronger than the forces of gravity. This can be visualized by imagining the electromagnets of cranes' in a junkyard holding up a complete car wreck before dropping it in the crushing mechanism. The electromagnetic forces overcome the gravitational force of the whole Earth on the car.

There are other interesting differences, like the fact that gravity is always an attractive force while electromagnetic forces may be either attractive and repulsive. Furthermore, the force of gravity cannot be shielded, it acts through any kind of barrier between the masses. On the other hand, electromagnetic forces can easily be shielded by certain insulating materials.

The similarities and differences are so inviting to attempt to reconcile the two classes that Einstein himself spent the three decades following the invention of his gravity theory trying to do just that. He never succeeded and we are left with the doubt whether that could ever be possible if neither Einstein's nor Eddington's genius was not enough.

The other direction of potential reconciliation is that of gravity with the nuclear forces. Since nuclear forces are propagated by various components of atoms, it is reasonable to hypothesize the existence of a yet unknown elementary particle that would be the graviton.

It is hoped that the electromagnetic, nuclear and gravitational forces can be merged into a single field theory, called the Grand Unified Theory (GUT). In this theory nature would have three fundamental forces. This would not be the first time that something important in nature has a trichotomy. Three is the perfect number after all.

But the final and most enduring mystery of gravity is the way it is propagated through space. There is a school of thought about that which believes that gravitation is truly carried by the mysterious gravitons. But since there is no way to detect them, we simply don't know.

As Einstein was stumped by the reconciliation with electromagnetism, so was Newton with the topic of the progression of gravity. He also considered the possibility of ether playing the role. This was not unheard of during the two millennia since Aristotle first proposed its existence and as we just saw it above, it might have a second life in connection with dark matter.

Newton was so disturbed by not being able to find the method of gravity's propagation that he wrote these words, fitting to conclude our journey: "That one body may act upon another at a distance through a vacuum without the mediation of anything else, by and through which their action and force may be conveyed from one another, is to me so great an absurdity that, I believe, no man who has in philosophic matters a competent faculty of thinking could ever fall into it."

Epilogue

We arrived at the end of our search to unravel gravity's mysteries. We learned that the historical interpretations of antiquity gave way to a scientific approach by Newton. That approach, while still practical in everyday circumstances, was inaccurate in some cases such as Mercury's precession.

We learned that gravity is one of the fundamental forces of nature. It's effect is proportional to the mass of objects, but not their weights. The weight of an object is in fact the product of its mass and the acceleration of gravity. Gravity is the balancing force that keeps Moon in orbit around Earth, in turn Earth around the Sun and so on in the universe. Gravity's effect decreases by the square of the distance between the objects producing the gravity but apparently this law, albeit beautifully simple, is not accurate enough to describe some phenomena.

Gravity provided the means of measuring the constitution of landmasses of Earth and influences the motion of man-made satellites. Our celestial partner, Moon's gravity is the main cause of tides. Gravity also influences the path of light and results in many intriguing phenomena, such as black holes.

Einstein's revolutionary idea of curved space seems

to be in synchrony with all of these phenomena. The gravitational theory of Einstein, at one time touted to be the ultimate theory, however, also has shortcomings in explaining some of the remaining mysteries of gravity. The Pioneer anomaly and the lack of our ability to detect the mechanism manifesting the gravitational attraction show that we have not reached the ultimate understanding yet.

This book was intended for readers who wanted to know how gravity works. It is hoped that there were some answers provided in the above, but it is clear that there are many remaining questions. Gravity remains a persistent mystery of humanity and we can continue to hypothesize about it.

It is possible that gravity is simply the remnant of the cohesive force of the fabric of the original material of the universe. The dark matter (or ether) is just the filler in the extremely sparse regions of the extraordinary expansion of the universe during its evolution.

In a sense it is the touch of the hand that originally created and spread the material of the universe, and that may ultimately pull it together again.

Bibliography

[1] Darling, D.: Gravity's arc, Wiley and Sons, New Jersey, 2006

[2] Gamow, G.: Gravity, Doubleday, New York, 1962

[3] Isaacson, W.: Einstein: His life and universe, Simon & Schuster, New York, 2007

[4] Feynman, R. P.: Lectures on gravitation, California Institute of Technology, Pasadena, 1995

[5] Greene, B.: The elegant universe, Random House, New York, 2000

[6] Moffat, J.: Reinventing gravity, Harper Collins, New York, 2008

[7] Komzsik, L.: Wheels in the sky: Keep on turning, Trafford publishing, 2010

[8] Naess, A. and Anderson, J.: Galileo Galilei: when the world was still, Springer, Berlin, 2005

[9] Singh, S.: Big bang: The origin of universe, Harper Collins, New York, 2004

[10] Thorne, K. S.: Black holes and time warps: Einstein's outrageous legacy, Norton and Co., New York, 1994

[11] White, M.: Isaac Newton: The last sorcerer, Perseus, New York, 1998

[12] Whittaker, E.: A history of theories of aether and electricity, Nelson and Sons, New York, 1953

[13] Wood, F. J.: Tidal dynamics, Reidel Publishing, 1986